Timo Gottschalk

Markov Processes in Stochastic Modeling of Transport Phenomena

Timo Gottschalk

Markov Processes in Stochastic Modeling of Transport Phenomena

From Application to Theory

Südwestdeutscher Verlag für Hochschulschriften

Impressum/Imprint (nur für Deutschland/ only for Germany)
Bibliografische Information der Deutschen Nationalbibliothek: Die Deutsche Nationalbibliothek verzeichnet diese Publikation in der Deutschen Nationalbibliografie; detaillierte bibliografische Daten sind im Internet über http://dnb.d-nb.de abrufbar.
Alle in diesem Buch genannten Marken und Produktnamen unterliegen warenzeichen-, marken- oder patentrechtlichem Schutz bzw. sind Warenzeichen oder eingetragene Warenzeichen der jeweiligen Inhaber. Die Wiedergabe von Marken, Produktnamen, Gebrauchsnamen, Handelsnamen, Warenbezeichnungen u.s.w. in diesem Werk berechtigt auch ohne besondere Kennzeichnung nicht zu der Annahme, dass solche Namen im Sinne der Warenzeichen- und Markenschutzgesetzgebung als frei zu betrachten wären und daher von jedermann benutzt werden dürften.

Verlag: Südwestdeutscher Verlag für Hochschulschriften Aktiengesellschaft & Co. KG
Dudweiler Landstr. 99, 66123 Saarbrücken, Deutschland
Telefon +49 681 37 20 271-1, Telefax +49 681 37 20 271-0, Email: info@svh-verlag.de
Zugl.: Bochum, Ruhr-Universität, Diss., 2008

Herstellung in Deutschland:
Schaltungsdienst Lange o.H.G., Berlin
Books on Demand GmbH, Norderstedt
Reha GmbH, Saarbrücken
Amazon Distribution GmbH, Leipzig
ISBN: 978-3-8381-0568-0

Imprint (only for USA, GB)
Bibliographic information published by the Deutsche Nationalbibliothek: The Deutsche Nationalbibliothek lists this publication in the Deutsche Nationalbibliografie; detailed bibliographic data are available in the Internet at http://dnb.d-nb.de.
Any brand names and product names mentioned in this book are subject to trademark, brand or patent protection and are trademarks or registered trademarks of their respective holders. The use of brand names, product names, common names, trade names, product descriptions etc. even without a particular marking in this works is in no way to be construed to mean that such names may be regarded as unrestricted in respect of trademark and brand protection legislation and could thus be used by anyone.

Publisher:
Südwestdeutscher Verlag für Hochschulschriften Aktiengesellschaft & Co. KG
Dudweiler Landstr. 99, 66123 Saarbrücken, Germany
Phone +49 681 37 20 271-1, Fax +49 681 37 20 271-0, Email: info@svh-verlag.de

Copyright © 2009 by the author and Südwestdeutscher Verlag für Hochschulschriften Aktiengesellschaft & Co. KG and licensors
All rights reserved. Saarbrücken 2009

Printed in the U.S.A.
Printed in the U.K. by (see last page)
ISBN: 978-3-8381-0568-0

Contents

Introduction V

1 Stochastic Modeling 1

    1.1 Markov Processes 3

2 Danckwerts' Law 8

    2.1 Original Work 9

        2.1.1 Residence Time Distributions 10

        2.1.2 Danckwerts' Groundwork from 1953 11

        2.1.3 Gibilaro's Extension of Danckwerts' Law 12

        2.1.4 Examples 15

    2.2 Discrete Case 17

I

		2.2.1	The Main Theorem	19

       2.2.2 Proof of the Main Theorem 20

       2.2.3 Further Results and Discussion of Theorem 2.1 21

       2.2.4 Examples 26

2.3 Continuous Case 32

       2.3.1 Diffusions 32

       2.3.2 Diffusions and Semigroup Theory 34

       2.3.3 Main Result and Further Structure of this Chapter 38

       2.3.4 Kolmogorov's Backward Equation with Boundary Conditions 39

       2.3.5 Conditions 42

       2.3.6 Deriving the Semigroup 46

       2.3.7 Proof of Theorem 2.3 61

       2.3.8 Examples 62

3 Multiphase Processes 64

3.1 Bubbling Fluidized Bed Reactor Revisited 65

    3.1.1 Original Model 66

    3.1.2 Multiphase Model 67

3.2 Danckwerts' Law Revisited 70

    3.2.1 Danckwerts' Law for the Twophase Fluidized Bed Reactor Model . 71

3.3 Diffusion Approximation 72

    3.3.1 The Processes 72

    3.3.2 Approximation 78

3.4 Partial Differential Equation for the Diffusion Limit 97

    3.4.1 Interior 99

    3.4.2 Boundaries 100

    3.4.3 Results 102

4 Slugging Fluidized Beds — 103

4.1 Slugging Fluidized Beds — 104

4.2 A Markov Chain Model for a Slugging Fluidized Bed — 105

4.2.1 Model Setup — 105

4.2.2 The Markov Chain Model — 108

4.3 Comparison of Model and Experiment — 113

4.3.1 Experimental Setup — 113

4.3.2 Model Setup — 114

4.3.3 Results — 115

Bibliography — 119

Introduction

Often theory flourishes when it is driven or inspired by an underlying certain need in applications. In the mathematical context the development of calculus by Gottfried W. Leibniz and Isaac Newton in the seventeenth century constitutes an illuminating example. There the need to precisely describe laws of nature in physics and being able to logically combine existing ones to find or generate others yielded this development. Nowadays calculus is fundamental to any mathematical education and the basis for many areas in present research. It has found its way not only into other sciences as physics, engineering, economics, psychology and many many more but also into the schools and can be viewed as an essential part of human education.

From a probabilist's point of view Albert Einstein's article in 1905 'Über die von der molekularkinetischen Theorie der Wärme geforderte Bewegung von in ruhenden Flüssigkeiten suspendierten Teilchen' which led to the discovery of Brownian motion in mathematics stands out. It is thanks to Norbert Wiener's work in 1923 that a rigorous mathematical foundation of Brownian motion by the Wiener process has been established. Without the Wiener process many modern methods in science and economics would be impossible, e.g. the precise description of the movement of a droplet of ink in water or the risk management for banks and insurance companies.

The aim of this work is to continue the portrayed procedure for several specific applications which arise in process technology. At the beginning of each section a concrete application, example or model is given which serves as a stimulus for the mathematical theory shown thereafter. As indicated in the aforementioned examples of calculus and Brownian motion the developed theory has a much wider application range as the motivational application itself and also promotes the development of mathematical theory.

The first chapter is an exception of this rule since it does not cover a specific stochastic model but it discusses the principles of stochastic modeling and presents the necessary mathematical objects. The notions of stochastic processes, especially Markov chains and more general Markov processes, are introduced. Properties and ways of describing and of analyzing of these processes are given. Next to a rigorous mathematical presentation emphasis is laid upon the interpretations of these mathematical objects and properties in familiar terms to further an intuitive understanding and applicability of these objects.

A law of nature is the originating point of Chapter 2. In 1953 Peter V. Danckwerts discovered the today called Danckwerts' law which states that the mean residence time of a fluid element in a continuous operated vessel with continuous and steady throughflow is equal to the volumetric in- or outflow velocity divided by the volume of the vessel. Since more often than not the behavior of particles or fluids in some processing vessel is hard to describe and to analyze it is even harder to ensure the validity of Danckwerts' law in the chosen model. In Chapter 2 two stochastic models which are easily and broadly applicable are discussed and conditions are given such that these models obey Danckwerts' law. But before it Danckwerts' original work and an extension by Gibilaro (1979) which is covered as well are presented.

In the first section of Chapter 2 a homogeneous Markov chain $(X_n)_{n \in \mathbb{N}}$ on the discrete state space $\{1, 2, \ldots, N+1\}$ is examined. Danckwerts' law is reformulated in mathematical language as

$$E[T|X_0 = 1] = \frac{v}{N}$$

for some stopping time T and some velocity parameter v. Afterwards the conditions which guarantee the validity of Danckwerts' law are introduced and shown to be sufficient. Further consequences of the imposed conditions are derived as well. These shed extra light on the interpretation of these conditions. Applications of the Markov chain model with its conditions to several reasonable examples conclude this section. Especially a very successful application of this model to a particular difficult to model system, a fluidized bed reactor, is given. Results from this section with a stronger focus on application can also be found in Gottschalk, Dehling and Hoffmann (2006) and Dehling, Gottschalk and Hoffmann (2007).

The second section of Chapter 2 is devoted to the continuous case. There a diffusion process $(X_t)_{t \in [0,1]}$ on $[0,1]$ is the object of interest. Thus some background on diffusions is given. To use the theory of strongly continuous semigroups in this context it is elaborated upon their relationsship to diffusion processes which has been discovered by William Feller around 1950. Thereafter a formulation of Danckwerts' law in the continuous case is given as

$$E\left[\lambda^1(\{t : X_t \in [0,1[\})|\ X_0 = 0\right] = \frac{1}{v_1}$$

where λ^1 denotes the one–dimensional Lebesgue measure and v_1 some velocity parameter. In comparison to the discrete case the analysis of the continuous case is of much higher complexity and needs stronger tools. Nevertheless the derivation and formulation of the conditions that ensure that Danckwerts' law holds and a proof of their sufficiency are presented. To this aim a translation of the stochastic problem into semigroup language is performed and the associated generator, its inverse and the corresponding semigroup are found. With these objects a short proof is possible. The explicit derivation of the inverse of the generator of the semigroup plays a crucial role in the proof. On the way further properties and characterizations of the diffusion process are deduced.

Chapter 3 deals with a particular class of Markov processes. These processes are called multiphase processes and are Markov processes with a state space of the form

$$S \times \{1, 2, \ldots, K\}$$

for some space S, e.g. $\{1, 2, \ldots, N\}$ or $[0, 1]$, and $K \geq 2$. The further investigation of the example of a fluidized bed reactor that has already been discussed before shows where these kind of processes arise naturally. With Chapter 2 in mind the validity of Danckwerts' law for multiphase processes is examined. Fortunately it can be shown that the results from Chapter 2 are easily extended to multiphase processes. The main contribution of Chapter 3 consists of an approximation theorem for multiphase processes. It is shown that a discrete multiphase process $(Y_t^N)_{t \in [0,1]}$ with sensible chosen transition probabilities weakly converges to a continuous multiphase process $(Y_t)_{t \in [0,1]}$, i.e.

$$(Y_t^N)_{t \in [0,1]} \xrightarrow{\mathcal{D}} (Y_t)_{t \in [0,1]}$$

holds for $N \to \infty$. Ideas by André R. Dabrowski and Herold G. Dehling who treated a simpler case of weak convergence of a Markov chain to a limiting Markov process in 1998 could be extended to multiphase processes. Moreover, under certain assumptions a set of partial differential equations with boundary conditions that characterizes the probability density of the limiting multiphase process is derived. The notion of a multiphase process, a discussion of Danckwerts' law for multiphase processes and an application of a multiphase process including successful model validation with empirical data are topics also treated in Dehling, Gottschalk and Hoffmann (2007).

Up to now only homogeneous Markov processes have been studied. This changes in Chapter 4. There a heterogeneous Markov chain as a model for transport in slugging fluidized beds is considered. At first an exposition on slugging fluidized beds is given and it is found that these special kind of systems are known to be notoriously difficult to model. Even though, a reasonable stochastic model for this system is introduced by using a heterogeneous Markov chain. The derivation of this model occupies a significant portion of Chapter 4 and features the translation from a physical setup into a mathematical one. Finally a validation of the model by different sets of experimental data is performed. Comparison of results generated with the considered Markov chain model with the empirical data indicates good applicability. A more extensive and applied study of these models is given in Dehling, Dechsiri, Gottschalk, Wright and Hoffmann (2006).

At this point I like to thank everybody who supported the genesis of this work, especially Prof. Dr. Herold Dehling.

Chapter 1

Stochastic Modeling

Understanding and predicting phenomena from nature, society, economics etc. has always been an important issue for mankind. Science evolved by abstracting from the matter at hand and pouring its essence into a model. Based upon this abstraction theories and hypotheses have been created and natural laws and other universally valid principles have been discovered. The huge impact of these discoveries has pressed this development ahead and even accelerated its pace. It has at first become clear in the natural sciences, that a rigorous and precise language is needed to perform this abstraction, i.e. to formulate the necessary models. This rigorous and precise language could be found in mathematics. As early as 1623 Galileo Galilei is supposed to have said: 'Nature is a book written in the language of mathematics'.

Nowadays it is common procedure for natural scientists to use a mathematical framework for their models. These mathematical frameworks or mathematical models can roughly be divided into two distinct categories, namely deterministic and stochastic models. Deterministic models are usually formulated via a partial differential equation derived from natural laws for some given quantities. Stochastic models describe systems involving some randomness or chance and focus onto some randomly behaving quantity. Discovering laws that rule the exhibited randomness of a system is the aim of the stochastic approach. Even though deterministic and stochastic modeling seem to be opposite of each other, relations between them exist. For instance in Section 2.3 the one-to-one correspondenc of certain Markov processes and strongly continuous semigroups is established.

The outcome of an experiment involving chance or the state of a stochastic system is usually denoted by a random variable X. A random variable $X : \Omega \to S$ is a measurable function defined on a probability space $(\Omega, \mathcal{F}, P)$ consisting of some set Ω, a σ-algebra $\mathcal{F}$ on Ω and a probability measure P on $\mathcal{F}$. The space S is called the *state space* of X and equipped with some σ-algebra $\mathcal{S}$. The distribution of X is then given by $P_X(A) = (P \circ X^{-1})(A) = P(X \in A)$ for all $A \in \mathcal{S}$. In the following the state space S and its corresponding σ-algebra will either be a compact interval in the real numbers with the Borel-σ-algebra $\mathcal{B}$, the smallest σ-algebra that contains all open sets or a finite set $M = \{1, 2, 3, \ldots, N\}$ and the power set $\mathcal{P}(M)$. Since we are interested in the evolution of a stochastic system in time we consider a whole family of random variables a socalled stochastic process.

Definition 1.0.1 *Let Ω be a non-empty set, $\mathcal{F}$ a σ-Algebra in Ω and P a probability measure on $\mathcal{F}$. Let I be a non-empty set, $(S, \mathcal{S})$ a measurable space and $(X_t)_{t \in I}$ a collection of $\mathcal{F}$-measurable functions with values in its state space $(S, \mathcal{S})$. Then the probability space $(\Omega, \mathcal{F}, P)$ together with $(X_t)_{t \in I}$ is called a* **stochastic process**.
For each $\omega \in \Omega$ the mapping from I to S defined by $t \mapsto X_t(\omega)$ is called a **path** *of the process.*

The common interpretation of the index set I is that it denotes time. In most cases it is a subset of the positive numbers, i.e. $I \subseteq [0, \infty)$ or equal to the nonnegative integers, i.e. $I = \mathbb{N} = \{0, 1, 2, \ldots\}$. In the first case we speak of a continuous system (in time) and in the second case of a discrete system (in time). The evolution of a stochastic system can show various degrees of dependence in its structure ranging from (stochastic) independence of the family $(X_t)_{t \in I}$ to total depence, e.g. if $X_t = X_0 e^t$ for all $t \in [0, 1]$. In the following we focus on a particular dependence structure known as the Markov property. It models that the evolution of a stochastic system does only depend on the present and not on the past or future. This notion will be made precise in the following section.

1.1 Markov Processes

A Markov process is a stochastic process with a special kind of dependence structure that describes a process without memory. Its future only depends on the present with no influences from the past. To formulate a Markov process precisely we need the concept of conditional probability. For a real-valued, integrable random variable X defined on a probability space $(\Omega, \mathcal{F}, P)$ there exists uniquely within almost sure equality a $\mathcal{F}_0$-measurable, integrable function X_0 for every sub-σ-algebra $\mathcal{F}_0$ of $\mathcal{F}$ such that $\int_A X_0 \, dP = \int_A X \, dP$ for all $A \in \mathcal{F}_0$. The function X_0 is called the *conditional expectation* of X under the hypothesis $\mathcal{F}_0$ and is written as $E(X|\mathcal{F}_0) = X_0$. The conditional expectation of the indicator function $\mathbb{1}_A$ for the set $A \in \mathcal{F}$ is called the *conditional probability* of A under the hypothesis $\mathcal{F}_0$ and is denoted by $P(A|\mathcal{F}_0) = E(\mathbb{1}_A|\mathcal{F}_0)$. Now we are able to give the definition of a Markov process.

Definition 1.1.1 *A stochastic process* $(\Omega, \mathcal{F}, P, (X_t)_{t \in I})$ *with totally ordered index set* I *and state space* $(S, \mathcal{S})$ *is called a* **Markov process** *if*

$$P(X_t \in B | \sigma(X_\tau : \tau \leq s)) = P(X_t \in B | \sigma(X_s)) \tag{1.1}$$

holds almost surely for all $s < t$, $s, t \in I$ *and all* $B \in \mathcal{S}$.

The σ-algebras $\sigma(X_\tau : \tau \leq s)$ and $\sigma(X_s)$ above are defined as the smallest σ-algebras on Ω such that all X_τ for $\tau \leq s$ or X_s is measurable respectively. The fact that a Markov process describes a process without memory shows when the condition encoding information about the entire past $\sigma(X_\tau : \tau \leq s)$ changes to one only encoding information about the present $\sigma(X_s)$. Equation (1.1) is called the *Markov property*.

Special functions socalled Markov kernels give rise to Markov processes. For a measurable space $(S, \mathcal{S})$ a function p on $S \times \mathcal{S}$ is called a *Markov kernel* if and only if $p(x, \cdot)$ is a probability measure for all $x \in S$ and $p(\cdot, B)$ is measurable for all $B \in \mathcal{S}$. Sometimes we allow that $p(x, \cdot)$ is a positive measure with mass less than one. Then the process can 'die' at some time and vanish. Those Markov kernels are usually called Submarkov kernels and lead to Submarkov transition functions and Submarkov processes. For convenience we do not want to distinguish between Submarkov and Markov kernels in the following since it does not affect the presented theory. A collection of Markov kernels linked together in a sensible way is called a *Markov transition function* and yields a Markov process. The precise definition is the following.

Definition 1.1.2 *Let $(S, \mathcal{S})$ be a measurable space. A family of functions $(p_t)_{t \geq 0}$ on $S \times \mathcal{S}$ is called a **Markov transition function** if and only if the conditions*

(i) p_t *is a Markov kernel,*

(ii) $p_0(x, B) = \delta_x(B),$

(iii) $p_{t+s}(x, B) = \int_S p_s(y, B) p_t(x, dy)$

hold for all $(x, B) \in S \times \mathcal{S}$ and $t, s \geq 0$.

To any given Markov transition function there exists a specific Markov process as shown in Theorem 1.1. Condition (iii) in the definition above

$$p_{t+s}(x, B) = \int_S p_s(y, B) p_t(x, dy)$$

is called the **Chapman-Kolmogorov equation**. This is the crucial ingredient that guarantees the Markovian structure of the stochastic process derived from $(p_t)_{t \geq 0}$. It reveals that the knowledge of the Markov kernels p_t in a small environment of zero determines completely the distribution of the corresponding stochastic process for all times $t \geq 0$. If one interprets $p_t(x, B)$ as the probability that a moving particle starting at a point x will be in the set B after time t, then the Chapman-Kolmogorov equation expresses that its movement only depends on its present position but not on its previous path.

For a Markov transition function $(p_t)_{t \geq 0}$, a probability measure μ on the measurable space $(S, \mathcal{S})$, $n \geq 1$, a subset J of $[0, \infty)$ of the form $J = \{t_1 < ... < t_n\}$ and every $B \in \mathcal{S}$ define finite-dimensional distributions via

$$P_J(B) := \int \int ... \int \mathbb{1}_B((x_1, ..., x_n)) \, p_{t_n - t_{n-1}}(x_{n-1}, dx_n) ... p_{t_1}(x_0, dx_1) \mu(dx_0) \, . \quad (1.2)$$

The starting distribution is given by the measure μ and the Markov kernels $p_{t_i - t_{i-1}}$ govern the transitions between the timesteps t_{i-1} and t_i. The existence of a Markov process with these finite-dimensional distributions is subject of the next theorem.

Theorem 1.1 *Let $(p_t)_{t\geq 0}$ be a Markov transition function, μ a probability measure on the measurable space $(S, \mathcal{S})$ and S a complete metric space with a countable basis. Then there exists a Markov process with state space $(S, \mathcal{S})$ and index set $[0, \infty)$ such that the finite-dimensional distributions are given via the Markov transition function as defined in Equation (1.2). Moreover,*

$$P(X_t \in B | \sigma(X_\tau : \tau \leq s)) = p_{t-s}(X_s, B)$$

holds almost surely for all $B \in \mathcal{S}$ and $0 \leq s < t$.

Proof. A proof of this theorem can be found in Bauer (1996). □

Another criteria to distinguish stochastic processes is the nature of their state spaces and index sets. Particularly if they are continuous or discrete, i.e. finite or countable. Methods, techniques and linguistic usage often differ for discrete and continuous processes. As we have seen before Markov transition functions and their associated Markov processes are only defined for the continuous index set $[0, \infty)$. For discrete Markov processes socalled Markov chains we like to present some theoretical background.

Markov Chains

A Markov process with finite state space S and index set $\mathbb{N}$ is called a *Markov chain* and denoted by $(X_n)_{n\geq 0}$. We assume a state space of the form $S = \{1, 2, \ldots, N\}$. A *starting measure* $\pi = (\pi_i)_{1 \leq i \leq N}$ is defined by $\pi_i = P(X_0 = i)$. The *(one step) transition probabilities* $p_{ij}(n) = P(X_{n+1} = j | X_N = i)$ form the *transition matrix* $P(n) = (p_{ij}(n))_{1 \leq i,j \leq N}$ and completely determine together with the starting distribution the distribution of the process since

$$P(X_n = j) = \left(\pi^t \left(\prod_{m=1}^{n} P(m) \right) \right)_j \tag{1.3}$$

holds for all $j \in S$. This is a consequence of the *Chapman-Kolmogorov equation*

$$P(X_n = k | X_l = i) = \sum_{j=1}^{N} P(X_m = j | X_l = i) P(X_n = k | X_m = j)$$

valid for all $0 \leq l < m < n$ and $1 \leq k, i \leq N$. A discrete stochastic process possesses the Markov property when

$$P(X_{n+1} = i_{n+1} | X_n = i_n, \ldots, X_0 = i_0) = P(X_{n+1} = i_{n+1} | X_n = i_n) \tag{1.4}$$

holds for all $i_{n+1}, \ldots, i_0 \in S$ and $n \geq 0$. Thus Equation (1.4) constitutes an equivalent representation of the Markov property for Markov chains. Special but very useful and common Markov chains possess transition probabilities that do not depend on time. Those are called *time homogeneous* or just *homogeneous*. For these Equation (1.3) becomes

$$P(X_n = j) = \left(\pi^t P^n\right)_j$$

where $P = (p_{ij})_{1 \leq i,j \leq N}$ denotes the homogeneous transition matrix.

Invariant Measures for Markov Chains

Often long-term behavior of a system is of interest. The questions are: 'Does the $\lim_{n \to \infty} P(X_n = i)$ exist?', 'Is it unique?' and 'How can it be calculated?'. In the following we consider only homogeneous Markov chains. The simple calculation

$$\begin{aligned} \lim_{n \to \infty} P(X_n = i) &= \lim_{n \to \infty} P(X_{n+1} = i) \\ &= \lim_{n \to \infty} \sum_{j=1}^N P(X_n = j) p_{ij} \\ &= \sum_{j=1}^N \left(\lim_{n \to \infty} P(X_n = j)\right) p_{ij} \end{aligned}$$

appears as

$$\varrho^t = \varrho^t P \tag{1.5}$$

for $\varrho = (\varrho_i)_{1 \leq i \leq N}$ with entries $\varrho_i = \lim_{n \to \infty} P(X_n = i)$ in matrix notation if the ϱ_i exist. Equation (1.5) states that the existence of a left eigenvector with nonnegative entries summing up to one for the transition matrix P is necessary for the existence of a limiting distribution. Such a vector is called an *invariant distribution*. To formulate sufficient conditions which ensure the existence and uniqueness of $\lim_{n \to \infty} P(X_n = i)$ we introduce aperiodicity and irreducibility of Markov chains. A Markov chain is called *aperiodic* if and only if the greatest common divisor of the set $\{n \geq 1 : P(X_n = i | X_0 = i) > 0\}$ is equal to one for all states $1 \leq i \leq N$. It is called *irreducible* if and only if there exists for all $1 \leq i,j \leq N$ a $n(i,j) > 0$ such that $P(X_{n(i,j)} = j | X_0 = i) > 0$. Irreducibility of a Markov chain means that each state $j \in S$ can be reached from each other state $i \in S$. The well-known theorem about the existence of a limiting distribution reads as follows.

Theorem 1.2 *Let $(X_n)_{n \geq 0}$ be a aperiodic and irreducible Markov chain with state space $\{1, 2, \ldots, N\}$. Then there exists a unique invariant distribution ϱ such that*

$$\lim_{n \to \infty} P(X_n = i) = \varrho_i$$

for all $1 \leq i \leq N$.

Proof. A proof can be found in Krengel (2002). □

Chapter 2

Danckwerts' Law

Natural laws form the base of natural sciences. Their discovery is the main task of natural science. From time to time new hithero unknown natural laws can be discovered and have to be implemented in existing theory or even force new theory to be developed to replace an old one. Sensible theory cannot ignore natural laws. The current chapter is devoted to one particular natural law and its implementation in theory.

For a long time stochastic models have been created to model transport phenomena. Roughly 55 years ago P.V. Danckwerts discovered a natural law which is these days named after him. Danckwerts' law first appeared in the journal 'Chemical Engineering Science' in 1953. Until today stochastic models where Danckwerts' law applies to are created without implementing it or ensuring its validity for different reasons including ignorance, unawareness or inability. The present chapter elucidates Danckwerts' law and explains its implementation when using discrete or continuous Markov processes.

Section 2.1 presents Danckwerts' law as originally publicated by Danckwerts himself and an extension by Gibilaro in 1979. In section 2.2 a discrete Markov chain model is set up in a very general framework. Conditions for this Markov chain are derived which ensure the validity of Danckwerts' law and the later extension by Gibilaro. Some discussion and further results as well as up to date examples are given. The last part of this chapter Section 2.3 deals with Danckwerts' law in the continuous case. Again conditions are given which guarantee the validity of Danckwerts' law. For the proof a relation between certain Markov processes and strongly continuous semigroups is established following an outline of the utilized theory of strongly continuous semigroups.

2.1 Original Work

There are multiple approaches when investigating transport phenomena. Best would be to derive explicitly the functional connection between any explanatory variables on the system's evolution in time and space. Often these functional connections cannot be calculated explicitly. For instance the system might be modeled such that the solution of a partial differential equation describes the sought-after functional connection. But this solution is unknown and cannot be computed. Or the system exhibits random behavior and this causes the evolution of the system to depend on chance. In this case it might be possible to calculate probability distributions in a stochastic setting.

When recovering of the complete functional connection fails certain key quantities of the system's behavior become even more valuable to know. Examples of these are eigenvalues which might determine long-term behavior and rate of convergence to steady states or expected values and variances of variables describing residence time in the vessel. The latter are in the focus of this chapter.

An object, we might think of a small particle, an atom, a molecule or a droplet of water, moves in a given enviroment, maybe some matter, in water, inside of a reactor or through the air after entering at some entrance and might leave it at some time through an exit or not. A path that is taken inside the environment during some time could depend on chance, i.e. a probability distribution. A meaningful quantity for this system is given by the *mean residence time* of the object inside the environment. Danckwerts derived a formula for the mean residence time of a fluid element in a vessel and Gibilaro extended this formula to subregions.

2.1.1 Residence Time Distributions

Consider a chemical processing vessel particulary a reactor, that is operated continuously with steady and continuous inflow and outflow of particles. The residence time of a particle is the length of the period that the particle spends inside the reactor, i.e. the time that elapses between entering and leaving the reactor. In general, the residence time will not be the same for all particles and can be regarded as the outcome of a non-negative random variable T. The distribution of T is then called the residence time distribution (RTD), and its distribution function $F(t) := P(T \leq t)$ is called the RTD-curve.

In the reactor particles undergo some treatment whose effect depends crucially on the amount of time the particles spend inside the reactor. Thus the residence time distribution is an extremely important characteristic of any chemical reactor and therefore the object of intense study in process technology literature (see e.g. Westerterp, van Swaaij and Beenackers (1987)). In Figure 2.1 we have shown some typical RTD-curves.

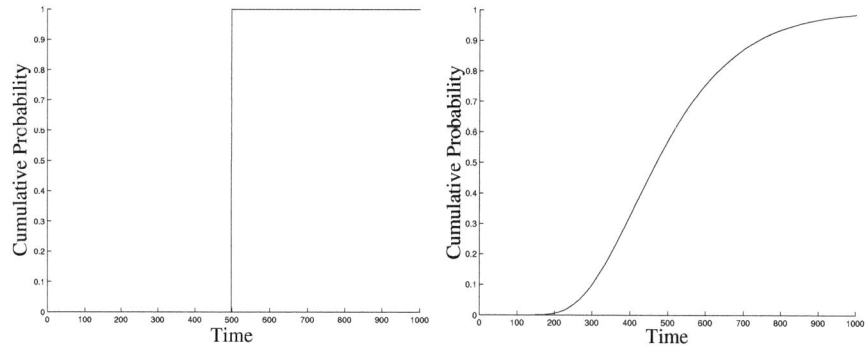

Figure 2.1: RTD-curves for 2 different Markov chain models

Both Markov chain models in Figure 2.1 have $N = 50$ cells and an in-/ and outflow constant $v = 0.1$. The transition probabilities for the figure on the left are set to $p_{ii+1} = 1$ for all $1 \leq i \leq N$. Those for the model on the right are given by

$$p_{ii+1} = 0.5,$$

$$p_{ii-1} = 0.4,$$

$$p_{ii} = 0.1$$

for $2 \leq i \leq N$ and

$$p_{11} = 0.5,$$
$$p_{12} = 0.5,$$
$$p_{5151} = 1$$

at the boundaries. This illustrates that the shapes of the RTD-curves depend heavily on various operating characteristics of a reactor and may vary enormously even among reactors with the same inflow/outflow rate. Theoretical prediction of the RTD-curve is a difficult task. In contrast to this, there is a simple formula for the mean residence time, discovered by Dankwerts in 1953. Danckwerts' law states that the mean residence time of a particle in a reactor with steady and continuous inflow/outflow equals the ratio of the volume of the reactor and its volumetric inflow/outflow.

2.1.2 Danckwerts' Groundwork from 1953

Danckwerts considers a processing vessel with continuous and steady throughflow. The objects of interest are the residence time distribution and its mean of fluid elements flowing through the vessel. The vessel has volume V and a volumetrically constant inflow of Q. Mass balance considerations lead to the conclusion that the outflow has to equal the inflow. In a thought experiment the vessel is supposed to be filled with white material before the inflow and only red material is added via the inflow. Thus the material in the vessel changes its color gradually from white to red.

The inflow starts at time $t = 0$. A function $F : [0, \infty) \to [0, 1]$ is defined such that its value $F(t)$ denotes the fraction of red material in the outflow at time t and can be interpreted as the distribution function of a random variable T describing the time to leave the vessel, i.e. $P(T \leq t) = F(t)$. This random variable is supposed to have a probability density $E : [0, \infty) \to [0, \infty)$, i.e.

$$F(t) = \int_0^t E(s)ds$$

holds for all $t \geq 0$. This probability density is called the *exit age*. Further another probability density $I : [0, \infty) \to [0, \infty)$ is introduced the socalled *internal age*. If

$G : [0, \infty) \to [0, 1]$ describes the fraction of red material inside the vessel at time t, it has probability density I, i.e.

$$G(t) = \int_0^t I(s)ds$$

holds for all $t \geq 0$. A mass balance consideration relates the quantities above. The amount of inflow until time t equals the red material inside the vessel at time t plus the red material that has left the vessel until time t. So

$$Qt = V \int_0^t I(s)ds + Q \int_0^t F(s)ds \qquad (2.1)$$

holds for all $t \geq 0$. Differentiating Equation (2.1) yields

$$Q = VI(t) + QF(t)$$
$$\Leftrightarrow 1 - F(t) = \tfrac{V}{Q} I(t)$$

for all $t \geq 0$. Now the mean residence time of the fluid elements can be computed as

$$<t> = \int_0^\infty tE(t)dt = \int_0^\infty 1 - F(t)dt = \int_0^\infty \frac{V}{Q} I(t) dt = \frac{V}{Q} \qquad (2.2)$$

since I is a probability density. Here $<t>$ denotes the expected value of the random variable T and can be interpreted as the mean residence time. Therefore **Danckwerts' law** states that the mean residence time of a fluid element in a vessel with continuous and steady throughflow and volumetrically constant inflow Q equals volume of the vessel V divided by inflow in the vessel.

2.1.3 Gibilaro's Extension of Danckwerts' Law

Gibilaro extended Danckwerts' law by showing that it is valid for any subregion inside the processing vessel. He also considered a vessel with continuous and steady throughflow and constant in- and outflow Q. In a subregion S of volume V_S a chemical reaction takes place transforming component A with constant ratio $k > 0$ into component B

$$A \xrightarrow{k} B.$$

Thus the concentration $C_A(t)$ of the component A in a fluid element that has resided in S for time t suffices

$$-\frac{d}{dt}C_A(t) = kC_A(t)$$

for all $t \geq 0$ and we obtain

$$C_A(t) = C_0 e^{-kt} \tag{2.3}$$

for all $t \geq 0$. The constant C_0 describes the fraction of material of component A of a fluid element that has never been in the subregion S, i.e. $C_A(0)$ and is therefore equal to the concentration of material of component A in the inflow $C_{A,in}$ which is assumed to be constant, i.e.

$$C_0 = C_A(0) = C_{A,in}. \tag{2.4}$$

Similar to Danckwerts Gibilaro introduces the exit age E_S being the probability density of a random variable modeling the time a fluid element has spend in the considered subregion S before leaving the vessel. Then the mean concentration of the component A in the outflow $C_{A,out}$ is given as

$$<C_{A,out}> = \int_0^\infty C_A(t) \, E_S(t) dt \tag{2.5}$$

where $<C_{A,out}>$ depends on the reaction rate k. Using Equations (2.3) and (2.4) and differentiating yields

$$\frac{d}{dk} <C_{A,out}> = -C_{A,in} \int_0^\infty t E_S(t) e^{-kt} dt.$$

The mean residence time in the subregion can now be obtained by taking the limit $k \to 0$

$$-C_{A,in}^{-1} \lim_{k \to 0} \frac{d}{dk} <C_{A,out}> = \int_0^\infty t E_S(t) dt = <t> \tag{2.6}$$

after exchanging integration and limit on the right hand side. Thus we need to compute the left hand side of Equation (2.6). Again analogous to Danckwerts a mass balance consideration provides the origin. Gibilaro supposes that the system reaches a steady state, i.e. that the underlying stochastic process has an invariant distribution; a clear fact from a physical point of view. After the system has reached a steady state the total inflow of material of component A in the vessel has to equal the outflow of material of component A plus the disappearance through the reaction of material of component A in the subregion S. This fact is expressed by

$$QC_{A,in} = Q <C_{A,out}> + k \int_S <C_A> dh \tag{2.7}$$

where $<C_A>$ denotes the mean fraction of material of component A in the subregion in the steady state. Differenting this equation and rearranging gives

$$\frac{d}{dk}<C_{A,out}> = -\frac{1}{Q}\left(\int_S <C_A> dh + k\left(\frac{d}{dk}\int_S <C_A> dh\right)\right)$$

since the left hand side of Equation (2.7) vanishes. Taking the limit for $k \to 0$ implies $<C_A> \to C_{A,in}$ and therefore

$$\lim_{k \to 0}\frac{d}{dk}<C_{A,out}> = -\frac{1}{Q}\int_S C_{A,in}\ dh = -C_{A,in}\frac{V_S}{Q}$$

after exchanging limit and integration. Inserting this into Equation (2.6) finally leads to the desired result

$$<t> = \frac{V_S}{Q}.$$

For any subregion in a vessel with continuous and steady throughflow and volumetrically constant inflow the mean residence time of a fluid element in this subregion is given by the quotient of its volume and the volumetric inflow.

A Stochastic Approach to Gibilaro's Extension

Here we like to present a stochastic approach to Gibilaro's extension as in Section 2.1.3 above. Although it is merely a reformulation we think it might make things clearer and show some underlying structure. Let us consider the same situation as outlined in Section 2.1.3 and denote by T_S the random variable that presents the total time a fluid element spends in the subregion S before leaving the vessel. Assume this random variable has a probability density E_S, i.e.

$$P(T_S \leq t) = \int_0^t E_S(s)ds$$

holds for all $t \geq 0$. Then Equation (2.5) is formulated as follows

$$E[C_A(T_S)] = \int_0^\infty C_A(t)\ E_S(t)dt \qquad (2.8)$$

using the identity $<C_{A,out}> = E[C_A(T_S)]$ and the same notation as in Section 2.1.3. After utilizing Equations (2.3) and (2.4) Equation (2.8) becomes

$$E[C_A(T_S)] = \int_0^\infty C_A(t)\ E_S(t)dt = C_{A,in}\int_0^\infty e^{-kt}\ E_S(t)dt = C_{A,in}m_{T_S}(-k) \qquad (2.9)$$

where m_{T_S} denotes the moment generating function of T_S as a function in k. Then the expected value of T_S is given as the first derivative of its moment generating function at 0 paralleling Equation (2.6)

$$E[T_S] = \frac{d}{dk}m_{T_S}(0) = -C_{A,in}^{-1}\left.\frac{d}{dk}\right|_{k=0} E[C_A(T_S)].$$

The same observations as in Section 2.1.3 yield likewise

$$QC_{A,in} = QE[C_A(T_S)] + k\int_S <C_A> dh$$

and

$$\left.\frac{d}{dk}\right|_{k=0} E[C_A(T_S)] = -\frac{1}{Q}\int_S C_{A,in}\, dh = -C_{A,in}\frac{V_S}{Q}$$

finally leading to

$$E[T_S] = -C_{A,in}^{-1}\left.\frac{d}{dk}\right|_{k=0} E[C_A(T_S)] = -C_{A,in}^{-1}\left(-C_{A,in}\frac{V_S}{Q}\right) = \frac{V_S}{Q}$$

the desired result. We note that existence of all quantities above is implicitly been assumed.

2.1.4 Examples

In the following two examples for Danckwerts' law are presented. They are special in a sense because it is in both cases possible and not even hard to analytically calculate the mean residence time. Nevertheless both cases present scientifically relevant models that are used in practice.

Plug-flow

We consider a processing vessel with height $h > 0$ and a flow from top to bottom with constant velocity $v > 0$ with no dispersion. This constant flow with no dispersion is called plug-flow. A fluid element enters the vessel at the top and exits once it has reached the bottom. In this actually deterministic process the height of fluid element at time $t \geq 0$ can be modeled with a random variable

$$X_t = h - tv$$

for $t \in [0, \frac{h}{v}]$. The residence time of the fluid element in the vessel is then given by

$$T = \inf\{t \geq 0 : X_t = 0\} = \frac{h}{v}$$

with distribution function

$$F(t) = P(T \leq t) = \mathbb{1}_{[\frac{h}{v}, \infty)}(t)$$

for all $t \geq 0$ visualized in Figure 2.1 on the left.

Ideally Stirred Mixer

An ideally stirred mixer is characterized by the fact that any material entering it is instantly distributed uniformly over the whole vessel. Let V denote the volume of the vessel, Q the volumetrically constant inflow and T the exit time of a particle entering at time $t = 0$. Then the mean residence time of a particle in the vessel is given by the expected value $E[T]$ of T. Let X_t be a random variable that describes the location of the particle in the vessel. Thus a particle that has entered the vessel at time $s = 0$ leaves the vessel between time 0 and t if and only if the particle enters the outflow region $O(t)$ during that time and has not left the vessel before, i.e.

$$P(T \leq t) = \int_0^t P(X_s \in O(s))(1 - P(T \leq s))ds \tag{2.10}$$

holds for all $t \geq 0$. We note that the volume $V_{O(t)}$ of the outflow region $O(t)$ is constant and equals Q, i.e.

$$V_{O(t)} = Q \tag{2.11}$$

for all $t \geq 0$ since the mass inside the vessel is preserved. The probability of belonging to the outflow equals

$$P(X_s \in O(s)) = \frac{1}{V} \int_{O(s)} \mathbb{1}(x)dx = \frac{V_O(s)}{V} = \frac{Q}{V} \tag{2.12}$$

for all $s \geq 0$ because of the instant uniform distribution over the whole vessel and Equation (2.11). In particular it is constant over time, too. Inserting the identity from Equation (2.12) in Equation (2.10) yields

$$P(T \leq t) = \int_0^t \frac{Q}{V}(1 - P(T \leq s))ds \qquad (2.13)$$

for all $t \geq 0$. By differentiating Equation (2.13) we obtain the differential equation

$$\frac{d}{dt}P(T \leq t) = \frac{Q}{V}(1 - P(T \leq t))$$

with solution

$$P(T \leq t) = 1 - e^{-\frac{Q}{V}t}$$

since $P(T \leq 0) = 0$. We identify this as the distribution function of an exponentially distributed random variable with expected value $\frac{V}{Q}$ and conclude

$$E[T] = \frac{V}{Q}.$$

2.2 Discrete Case

In many applications it is advantageous to consider discrete models. These usually allow good modeling and interpretation. Especially simulating becomes easy. Therefore in this chapter we present a discrete model and further the conditions which have to be imposed for Danckwerts' law to hold. A model for the processing vessel, i.e. a reactor and a Markov chain describing the movement of a single particle are given. We state the main condition for Danckwerts' law to hold, show that it is sufficient and discuss its interpretation by deriving additional results.

We imagine the reactor to be split in finitely many alike cells, numbered $1, \ldots, N+1$, with entrance in cell 1 and exit in cell $N+1$. We model the path of a single particle via a homogeneous Markov chain $(X_n)_{n \geq 0}$ where $X_n \in \{1, \ldots, N+1\}$ denotes the position of the particle at time $n \in \mathbb{N}$. This Markov chain is fully specified by its transition probabilities $p_{ij} := P(X_n = j | X_{n-1} = i)$ $(1 \leq i, j \leq N+1, n \in \mathbb{N})$ and its starting distribution, i.e. $(P(X_0 = i))_{1 \leq i \leq N+1}$.

We assume that the particles' movements are independent of each other, that the particles are incompressible and that the reactor has no void parts. This implies that the number of particles in each cell is constant over time, say M. Since the number of particles in the reactor is large the Law of Large Numbers applies and thus the flow from cell i to cell j is given as $M \cdot p_{ij}$ and p_{ij} can be interpreted as the fraction of material that flows from cell i to cell j in one timestep. We picture the model as in Figure 2.2.

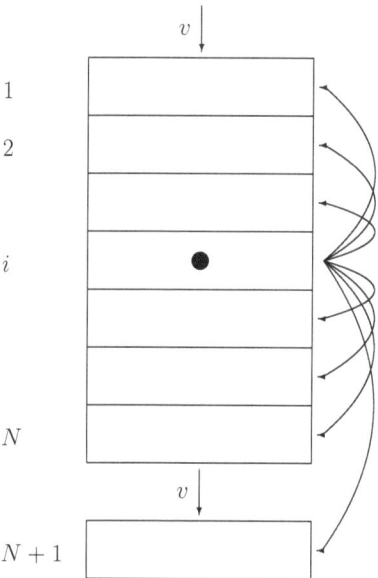

Figure 2.2: Reactor split in cells $1 \ldots N$ with exit in cell $N+1$, in-/outflow v, the particle in cell i and possible transition pathes to the right.

To translate Dankwerts' law into the language of Markov chains we define the first exit time of the Markov chain $T := \inf\{n \geq 0 : X_n = N+1\}$ and an inflow/outflow constant $v \in (0, 1]$ which describes the steady and continuous inflow and outflow. Now Danckwerts' law can be expressed as follows

$$E[T|X_0 = 1] = \frac{N}{v}. \tag{2.14}$$

2.2.1 The Main Theorem

For Danckwerts' law to hold some conditions on the transition probabilities have to be made. We want to move the discussion of these matters to section 2.2.3 and first present the main result.

Theorem 2.1 *Let $(X_n)_{n\geq 0}$ be a homogeneous Markov chain with transition probabilities p_{ij} sufficing $p_{N+1\,N+1} = 1$, $P(X_{n(i)} = N+1 | X_0 = i) > 0$ for some $n(i) \geq 0$ and all $1 \leq i \leq N$ and state space $\{1, 2, \ldots, N+1\}$. Let $v \in (0, 1]$ and let the following conditions*

(a) $\sum_{j=1}^{N} p_{j1} = 1 - v$,

(b) $\sum_{j=1}^{N} p_{ji} = 1$ for all $2 \leq i \leq N$

hold. Then the identity

$$E[T|X_0 = 1] = \frac{N}{v}$$

holds with $T = \inf\{n \geq 0 : X_n = N+1\}$. Further

$$E\left[\sum_{n=0}^{\infty} \mathbb{1}_{\{X_n = j\}} \Big| X_0 = 1\right] = \frac{1}{v}$$

holds for all $1 \leq j \leq N$.

Theorem 2.1 states that Danckwerts' law holds for Markov chains with mass balance conditions (a) and (b), inflow/outflow velocity v and no isolated states. Especially the expected residence time in any state $1 \leq j \leq N$ when starting in state 1 equals $\frac{1}{N}$. This implies that for every subdomain $S \subset \{1, 2, \ldots, N\}$ the expected residence time when starting in state 1 equals $\frac{V_S}{v}$ where V_S denotes the volume of the subdomain S. This is in fact the extension of Dankwerts' law from Gibilaro as presented in Section 2.1.3.

2.2.2 Proof of the Main Theorem

We first state some auxiliary results that will be needed in the proof of Theorem 2.1. These results are known (see e.g. Grinstead and Snell (1998)), nevertheless we present the proofs for the sake of completeness.

Theorem 2.2 *Let $(X_n)_{n\geq 0}$ be a homogeneous Markov chain with transition probabilities p_{ij} sufficing $p_{N+1\,N+1} = 1$ and $P(X_{n(i)} = N+1|X_0 = i) > 0$ for some $n(i) \geq 0$ and all $1 \leq i \leq N$ and state space $\{1, 2, \ldots, N+1\}$. Then the following holds.*

(i) The matrix $I - Q$ with I the identity matrix and $Q = (p_{ij})_{1\leq i,j \leq N}$ is invertible.

(ii) $(I - Q)^{-1} = \sum_{n=0}^{\infty} Q^n$

(iii) $(I - Q)^{-1}_{ij} = E\left[\sum_{n=0}^{\infty} \mathbb{1}_{\{X_n = j\}} \Big| X_0 = i\right]$

(iv) $E[T|X_0 = i] = \sum_{j=1}^{N} (I-Q)^{-1}_{ij}$ with $T = \inf\{n \geq 0 : X_n = N+1\}$

Proof. We note that for $P = (p_{ij})_{1\leq i,j\leq N+1}$ the n-times transition probabilities are given by $p_{ij}^{(n)} = P(X_n = j|X_0 = i) = (P^n)_{ij} = (Q^n)_{ij}$ for all $1 \leq i, j \leq N$. The last equality holds since the state $N+1$ is absorbing. Further $P(X_{n(i)} = N+1|X_0 = i) > 0$ for some $n(i) \geq 0$ and all $1 \leq i \leq N$ implies that there exists $n_0 \geq 0$ such that $\|Q^n\| < 1$ for all $n \geq n_0$ and for some norm $\|\cdot\|$. Therefore $\sum_{n=0}^{\infty} Q^n$ converges absolutely. The calculation

$$(I-Q)\sum_{n=0}^{\infty} Q^n = \left(\sum_{n=0}^{\infty} Q^n\right)(I-Q) = I$$

shows that $\sum_{n=0}^{\infty} Q^n = (I-Q)^{-1}$ holds thus proving (i) and (ii). The identities

$$E\left[\sum_{n=0}^{\infty} \mathbb{1}_{\{X_n = j\}} \Big| X_0 = i\right] = \sum_{n=0}^{\infty} (Q^n)_{ij} = \left(\sum_{n=0}^{\infty} Q^n\right)_{ij} = (I-Q)^{-1}_{ij}$$

for all $1 \leq i, j \leq N$ yield (iii). Since the time the chain being in the set of the nonabsorbing states equals the time to absorption, i.e. $T = \sum_{j=1}^{N}\sum_{n=0}^{\infty} \mathbb{1}_{\{X_n = j\}}$, we obtain (iv) by considering the identity

$$E[T|X_0 = i] = \sum_{j=1}^{N} E\left[\sum_{n=0}^{\infty} \mathbb{1}_{\{X_n = j\}} \Big| X_0 = i\right]$$

for all $1 \leq i \leq N$ and (iii). □

Now we are able to prove Theorem 2.1.

Proof of Theorem 2.1 Consider $P = (p_{ij})_{1 \leq i,j \leq N+1}$. Clearly Theorem 2.2 applies and therefore $I - Q$ with I the identity matrix and $Q = (p_{ij})_{1 \leq i,j \leq N}$ is invertible. Then the equation

$$(I - Q)^t x = e_1$$

where $e_1 = (1, 0, \cdots, 0)^t \in \mathbb{R}^n$ and $(I - Q)^t$ is the transposed matrix of $(I - Q)$ has a unique solution $x \in \mathbb{R}^n$ which is the first row of $(I - Q)^{-1}$. Since by conditions (a) and (b) the equalities $\sum_{i=1}^{N} (I - Q)_{1i}^t = v$ and $\sum_{i=1}^{N} (I - Q)_{ji}^t = 0$ $(2 \leq j \leq N)$ hold the vector $x_0 = \frac{1}{v}(1, \cdots, 1) \in \mathbb{R}^n$ solves the above equation and we obtain

$$(I - Q)^{-1} = \begin{pmatrix} x_0 \\ * \\ \cdot \\ \cdot \\ \cdot \\ * \end{pmatrix}.$$

This yields

$$E[T|X_0 = 1] = \sum_{i=1}^{N} \left((I - Q)^{-1}\right)_{1i} = \frac{N}{v}$$

by Theorem 2.2 (iv). Finally Theorem 2.2 (iii) implies $E\left[\sum_{n=0}^{\infty} \mathbb{1}_{\{X_n = j\}} \middle| X_0 = 1\right] = \frac{1}{v}$ for all $1 \leq j \leq N$. □

2.2.3 Further Results and Discussion of Theorem 2.1

Conditions (a) and (b) in Theorem 2.1 imply a third condition

(c) $\sum_{j=1}^{N} p_{jN+1} = v$

which can substitute condition (a). The following remark shows that a further generalization is possible.

Remark 2.2.1 Let $(p_{ij})_{1 \leq i,j \leq N+1}$ be the transition probabilities of a Markov chain with state space $\{1, 2, \ldots, N+1\}$. If any N of the following $N+1$ equations hold, then all $N+1$ equations hold.

(1) $\quad \sum_{j=1}^{N} p_{j1} = 1 - v$

(2...N) $\quad \sum_{j=1}^{N} p_{ji} = 1 \quad$ for all $2 \leq i \leq N$

(N+1) $\quad \sum_{j=1}^{N} p_{jN+1} = v$

Proof. Let all equations 1 to $N+1$ except for equation $1 \leq i_0 \leq N+1$ hold. The law of total probability yields

$$\begin{aligned} N &= \sum_{j=1}^{N} \sum_{i=1}^{N+1} p_{ji} \\ &= \sum_{j=1}^{N} \left(p_{ji_0} + \sum_{i=1, i \neq i_0}^{N+1} p_{ji} \right) \\ &= \sum_{j=1}^{N} p_{ji_0} + \sum_{i=1, i \neq i_0}^{N+1} \sum_{j=1}^{N} p_{ji}. \\ &=: (I_0) + (I) \end{aligned}$$

If $i_0 = 1$, then $(I) = N - 1 + v$ implies $(I_0) = 1 - v$, if $i_0 = N + 1$, then $(I) = N - v$ implies $(I_0) = v$ and if $2 \leq i_0 \leq N$ then $(I) = N - 2 + 1 - v + v = N - 1$ implies $(I_0) = 1$. □

We now want to discuss for a moment the physical meaning of the conditions imposed in Theorem 2.1. The inflow in the reactor is given as the outflow of the entrance, i.e. cell 1, minus the inflow in cell 1, i.e. $1 - \sum_{j=1}^{N} p_{j1} = v$ by condition (a) in Theorem 2.1. Particles that have left the reactor, i.e. reached the exit, i.e. cell $N+1$, cannot return. Thus the exit is an absorbing state, i.e. $p_{N+1N+1} = 1$. Further the outflow of the reactor is given as the inflow to the exit out of the reactor, i.e. $\sum_{j=1}^{N} p_{jN+1} = v$ by condition (c) above and equal to the inflow in the reactor.

In the interior of the reactor a mass balance equation between all cells has to hold since the particles are incompressible and the reactor has no void parts. This leads to condition (b) in Theorem 2.1, which will be further examined around Lemma 2.2.3. The absence of void parts in the reactor is implied in Theorem 2.1 by the condition that the absorbing state $N+1$ is reachable from every state. We will see that in Corollary 2.2.2. Finally we give an example of a Markov chain satisfying all conditions of Theorem 2.1 except for condition (b) and not sufficing Danckwerts' law.

We start by showing that the reactor has no void parts, i.e. all cells are reachable from the entrance.

Corollary 2.2.2 *Under the condition of Theorem 2.1 there is for every state $1 \leq i \leq N$ a positive probabiltity to be reached from state 1, i.e. $P(X_{n(i)} = i | X_0 = 1) > 0$ for some $n(i) \geq 0$.*

Proof. By Theorem 2.1 it holds that

$$0 < \frac{1}{v} = x_0(i) = E\left[\sum_{n=0}^{\infty} \mathbb{1}_{\{X_n = i\}} \middle| X_0 = 1\right]$$

for all $1 \leq i \leq N$. This implies that

$$P(X_{n(i)} = i | X_0 = 1) > 0 \quad \text{for some } n(i) \geq 0 \text{ and all } 1 \leq i \leq N.$$

□

We point out that assuming irreducibility of the Markov chain is too strong. That can be seen in Examples 2.2.4. Now we turn our attention to conditions (a) and (b) in Theorem 2.1. These mass balance equations in the interior (condition (b)) with a constant inflow of v (condition (a)) in the reactor are equivalent to some flow condition as shown in the following lemma.

Lemma 2.2.3 *Let $(X_n)_{n \geq 0}$ be a homogeneous Markov chain with transition probabilities p_{ij}, absorption at $N+1$ and state space $\{1, 2, \ldots, N+1\}$. Let $v \in [0, 1]$. Then the following two conditions are equivalent.*

(i) $\sum_{k=1}^{i} \sum_{j=i+1}^{N+1} (p_{kj} - p_{jk}) = v \quad \text{for all } 1 \leq i \leq N$

(ii) (a) $\sum_{j=1}^{N} p_{j1} = 1 - v$

(b) $\sum_{j=1}^{N} p_{ji} = 1 \text{ for all } 2 \leq i \leq N$

Proof. Assume that condition (i) holds. Having in mind that $p_{N+1\,N+1} = 1$ we obtain for $i = 1$

$$v = \sum_{j=2}^{N+1}(p_{1j} - p_{j1}) = \sum_{j=1}^{N+1} p_{1j} - \sum_{j=1}^{N+1} p_{j1} = 1 - \sum_{j=1}^{N} p_{j1}$$

which is equivalent to condition $(ii)\,(a)$. For $2 \leq i \leq N$ we calculate

$$\begin{aligned}
\sum_{j=1}^{N} p_{ji} &= N - \sum_{\substack{j=1 \\ j \neq i}}^{N+1} \sum_{k=1}^{N} p_{kj} \\
&= N - 1 + v - \sum_{j=2}^{i-1}\sum_{k=1}^{N} p_{kj} - \sum_{j=i+1}^{N+1}\sum_{k=1}^{N} p_{kj} \\
&= N - 1 + \sum_{k=1}^{i}\sum_{j=i+1}^{N+1}(p_{kj} - p_{jk}) - \sum_{j=2}^{i-1}\sum_{k=1}^{N} p_{kj} - \sum_{j=i+1}^{N+1}\sum_{k=1}^{N} p_{kj} \\
&= N - 1 - \sum_{j=2}^{i-1}\sum_{k=1}^{N} p_{kj} - \sum_{j=i+1}^{N+1}\sum_{k=1}^{i} p_{jk} - \sum_{j=i+1}^{N+1}\sum_{k=i+1}^{N} p_{kj} \\
&= N - 1 - \sum_{j=2}^{i-1}\sum_{k=1}^{N} p_{kj} - \sum_{j=i+1}^{N+1}\sum_{k=1}^{i} p_{jk} - \sum_{k=i+1}^{N}\left(1 - \sum_{j=1}^{i} p_{kj}\right) \\
&= i - 1 - \sum_{j=2}^{i-1}\sum_{k=1}^{N} p_{kj}.
\end{aligned}$$

This shows condition $(ii)\,(b)$ for $i = 2$ and then iteratively for all $2 \leq i \leq N$.

Assume now that condition (ii) holds. It follows that

$$\begin{aligned}
\sum_{j=i+1}^{N+1}\sum_{k=1}^{i}(p_{kj} - p_{jk}) &= \sum_{j=i+1}^{N+1}\sum_{k=1}^{i} p_{kj} - \sum_{k=1}^{i}\left(1 - \sum_{j=1}^{i} p_{jk}\right) + v \\
&= \sum_{k=1}^{i}\sum_{j=1}^{N+1} p_{kj} - i + v \\
&= v
\end{aligned}$$

holds for all $1 \leq i \leq N$. □

Condition (i) in the above lemma describes that the flow of particles from the upper cells downwards is constant with rate v through every membrane between two cells, i.e. the flow from all cells above cell $i+1$ downwards $\sum_{k=1}^{i}\sum_{j=i+1}^{N+1} p_{kj}$ minus the flow from all the cells below cell i upwards $\sum_{k=1}^{i}\sum_{j=i+1}^{N+1} p_{jk}$ equals v (see Figure 2.3). This is equivalent

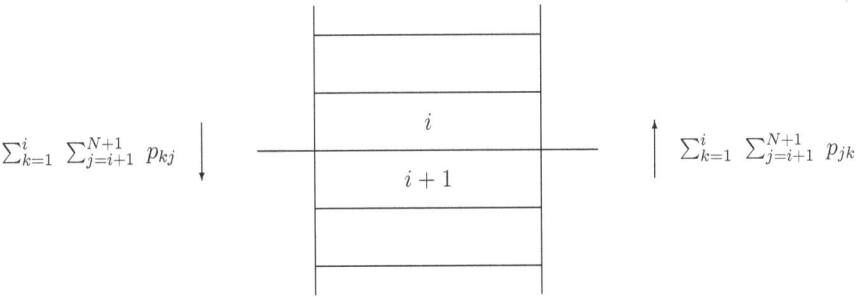

Figure 2.3: Illustration of the flow through a membrane between cells i and $i+1$.

to condition (ii) in the above lemma which describes that the inflow in the system from cell 1 which equals $1 - \sum_{j=1}^{N} p_{j1}$ is constant with rate v, the outflow of the system in the absorbing cell $N+1$ which equals $\sum_{j=1}^{N} p_{jN+1}$ is also constant with rate v and that the masses in the cells 2 to N do not change, i.e. $\sum_{j=1}^{N} p_{ji} = 1$.

The following example shows that Theorem 2.1 fails if condition (b) in Theorem 2.1 is violated even if all other conditions in Theorem 2.1 are met.

Example 2.2.4 Let $(X_n)_{n \geq 0}$ be a Markov chain with state space $\{1, 2, 3, 4\}$, $v, \varepsilon \in (0, 1)$ and transition probability matrix

$$P = \begin{pmatrix} 1-v & v & 0 & 0 \\ 0 & 1-\varepsilon & \varepsilon & 0 \\ 0 & 0 & 1-v & v \\ 0 & 0 & 0 & 1 \end{pmatrix}$$

Then $E[T|X_0 = 1] = \frac{2}{v} + \frac{1}{\varepsilon}$ with $T = \inf\{n \geq 0 : X_n = 4\}$.

Proof. After setting

$$Q = (p_{ij})_{1 \leq i,j \leq 3} = \begin{pmatrix} 1-v & v & 0 \\ 0 & 1-\varepsilon & \varepsilon \\ 0 & 0 & 1-v \end{pmatrix}$$

and calculating

$$(I-Q)^{-1} = \begin{pmatrix} v & -v & 0 \\ 0 & \varepsilon & -\varepsilon \\ 0 & 0 & v \end{pmatrix}^{-1} = \begin{pmatrix} \frac{1}{v} & \frac{1}{\varepsilon} & \frac{1}{v} \\ 0 & \frac{1}{\varepsilon} & \frac{1}{v} \\ 0 & 0 & \frac{1}{v} \end{pmatrix}$$

the claim follows via Theorem 2.2(iv). □

2.2.4 Examples

To further illustrate the results above we show some examples. We start by revisiting the examples from the previous Section 2.1.4 and discussing these in a discrete framework. In addition to this we present a very successful Markov chain model for the movement of a particle in a bubbling fluidized bed and its form after implementing the theory derived above. This model appeared first in Dehling et al. (1999).

Plug-flow Revisited

In the discrete framework plug-flow is characterized by its distinct transition probabilities. These are all equal to zero or to one showing the model's deterministic nature. But before considering the transition probabilities we model the vessel by a collection of cells $\{1, 2, \ldots, N, N+1\}$ visualized in Figure 2.2. The cell with the number 1 denotes the entrance and the cell $N+1$ denotes the exit, i.e. $p_{N+1N+1} = 1$. The transition probabilities are then set as follows

$$p_{ii+1} = 1$$

for all $1 \leq i \leq N$. Thus the Markov chain $(X_n)_{n \in \mathbb{N}}$ with starting probability $P(X_0 = i) = \delta_{1i}$ suffices

$$X_n = \min\{n+1, N+1\}$$

for all $n \in \mathbb{N}$ and yields for its first exit time $T = \inf\{n \geq 0 : X_n = N+1\}$

$$E[T] = N = \frac{N}{1}.$$

This corresponds to Danckwerts' law as formulated in Equation (2.14) with inflow constant $v = 1$. We note that the conditions for Danckwerts' law from Theorem 2.1

(a) $\sum_{j=1}^{N} p_{j1} = 0 \qquad = 1 - v,$

(b) $\sum_{j=1}^{N} p_{ji} = p_{i-1\,i} = 1 \qquad$ for all $2 \leq i \leq N$

are valid.

Ideally Stirred Mixer Revisited

As in the previous section we model the vessel by a collection of cells $\{1, 2, \ldots, N, N+1\}$, $N \geq 2$, with entrance from cell 1 and absorbing exit in cell $N+1$. We specify a Markov chain by postulating

$$P(X_0 = 1) = 1,$$

$$p_{ij} = \begin{cases} 1 - v & (i,j) \in \{1\} \times \{1\} \\ \frac{v}{N-1} & (i,j) \in \{1\} \times \{2, \ldots, N\} \\ 0 & (i,j) \in \{2, \ldots, N\} \times \{1\} \\ (1 - \frac{v}{N-1})\frac{1}{N-1} & (i,j) \in \{2, \ldots, N\} \times \{2, \ldots, N\} \\ \frac{v}{N-1} & (i,j) \in \{2, \ldots, N\} \times \{N+1\} \end{cases}$$

for some inflow $v \in (0, 1]$ and remembering that $p_{N+1\,i} = \delta_{N+1\,i}$ holds for the absorbing exit. This Markov chain simulates a system with an in- and outflow of v and completely random movement inside. If a particle has entered the system it cannot leave it again through the entrance but at each step through the exit with probability $\frac{v}{N-1}$. Once a particle has entered the interior of the vessel $\{2, \ldots, N\}$ at each transition one can imagine somebody throwing first a coin which shows heads with probability $\frac{v}{N-1}$ and tails with probability $1 - \frac{v}{N-1}$ and then either removing the particle from the vessel if a head was thrown or placing it randomly into another cell inside the reactor maybe determind by the throw of a fair dice with sides $2, \ldots, N$ if tails occured.

The random variable describing the residence time $T = \{n \geq 0 : X_n = N+1\}$ minus 2 can be regarded as the sum of two geometric random variables T_1 and T_2 with parameters v and $\frac{v}{N-1}$ respectively. The first random variable $T_1 + 1$ corresponds to the time the Markov chain stays in state 1, i.e. stays in the entrance, where at each time step the probability to leave the entrance into the vessel equals $P(X_{n+1} \in \{2, \ldots, N\} | X_n = 1) = \sum_{i=2}^{N} p_{1i} = \sum_{i=2}^{N} \frac{v}{N-1} = v$. Analogously the random variable $T_2 + 1$ corresponds to the time the Markov chain stays in the interior, i.e. in the states $\{2, \ldots, N\}$ where it has the probability to exit the vessel of $P(X_{n+1} = N+1 | X_n \in \{2, \ldots, N\}) = \frac{v}{N-1}$ at each timestep. Thus the expected value of T is given by

$$\begin{aligned} E[T] &= E[T_1] + E[T_2] + 2 \\ &= \frac{1-v}{v} + \frac{1-\frac{v}{N-1}}{\frac{v}{N-1}} + 2 \\ &= \frac{N-1-v}{v} + \frac{1-v}{v} + \frac{2v}{v} \\ &= \frac{N}{v} \end{aligned}$$

which suffices Danckwerts' law. Again we note that the conditions for Danckwerts' law from Theorem 2.1

(a) $\sum_{j=1}^{N} p_{j1} = p_{11} = 1 - v$,

(b) $\sum_{j=1}^{N} p_{ji} = \frac{v}{N-1} + \sum_{j=2}^{N}(1 - \frac{v}{N-1})\frac{1}{N-1} = 1 \qquad$ for all $2 \leq i \leq N$

are valid.

Bubbling Fluidized Bed Reactor

Here we present a Markov chain model for a particular complex system, namely a fluidized bed reactor. This model was first proposed by Hoffmann and Paarhuis (1990) and we discuss its improved form from Dehling et al. (1999). A fluidized bed reactor is a special kind of chemical reactor where gas is injected through a porous distributor plate at the bottom of the reactor. At a certain velocity of the gas the powder inside the reactor starts to float and exhibit a fluid-like behavior. This socalled fluidized bed is disturbed by bubbles forming above the distributor plate and rising to the top of the reactor. We suppose that this fluidized bed reactor is operated continuously, i.e. particles are added into the reactor at the top and removed at the bottom with constant and equal rates. A sketch of such a bubbling fluidized bed reactor is given in Figure 2.4.

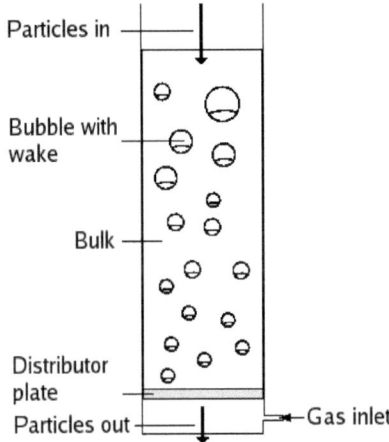

Figure 2.4: Bubbling fluidized bed reactor

Rowe and Partridge (1962) first postulated that the movement of a particle in a bubbling fluidized bed is governed by the following three main physical processes.

1. Transport downward in the bulk due to the removal of material at the bottom of the bed.
2. Dispersion due to the disturbance of the bulk material created by the rising bubbles.
3. Transport upward in bubble wakes and deposition on top of the bed.

According to these main processes a Markov chain model is set up. The reactor is split into N cells of equal size from the top cell number 1 down to the bottom cell N. An extra absorbing exit cell with the label $N+1$ is added. Therefore the state space of the Markov chain $(X_n)_{n\geq 0}$ is given by

$$S = \{1, 2, \ldots, N+1\}$$

such that X_n describes the height of a particle at time $n \in \mathbb{N}$. The transition matrix P has entries

$$\begin{aligned}
p_{i,i} &= \alpha_i = 1 - \beta_i - \delta_i - \lambda_i, \\
p_{i,i+1} &= \beta_i, \\
p_{i,i-1} &= \delta_i, \\
p_{i,1} &= \lambda_i
\end{aligned}$$

for $3 \leq i \leq N$,

$$\begin{aligned}
p_{2,1} &= \delta_2 + \lambda_2, \\
p_{2,2} &= \alpha_2, \\
p_{2,3} &= \beta_2,
\end{aligned}$$

for transitions originating in cell number 2 and

$$\begin{aligned}
p_{1,1} &= 1 - \beta_1, \\
p_{1,2} &= \beta_1, \\
p_{N+1,N+1} &= 1
\end{aligned}$$

at the boundaries.

These transition probabilities feature the required characteristics. The parameteres α_i, β_i and δ_i model a drift and dispersion in the bulk like an ordinary birth-death Markov chain. The parameter λ_i causes the extra transport of particles caught in the wake of rising gas bubbles to the top of the reactor via a jump. This setup is shown in Figure 2.5.

The mean residence time is usually not calcuable for this model but the theory introduced before gives conditions which ensure that Danckwerts' law holds in this model. These are

(a) $\sum_{j=1}^{N} p_{j1} = 1 - \beta_1 + \delta_2 + \sum_{j=2}^{N} \lambda_j = 1 - v,$

(b) $\sum_{j=1}^{N} p_{ji} = 1 - \delta_i - \beta_i - \lambda_i + \delta_{i+1} + \beta_{i-1} = 1 \qquad$ for all $2 \leq i \leq N$

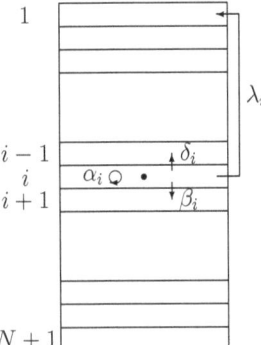

Figure 2.5: Discretized fluidized bed with arrows indicating all possible transition paths for the particle located in cell i.

and can be rewritten as

(a) $\beta_1 - \delta_2 - \sum_{j=2}^{N} \lambda_j = v,$

(b) $\beta_i - \delta_{i+1} = (\beta_{i-1} - \delta_i) - \lambda_i \quad$ for all $2 \leq i \leq N.$

Iterating in condition (b) now leads to

$$\beta_i - \delta_{i+1} = (\beta_1 - \delta_2) - \sum_{j=2}^{i} \lambda_j$$

for all $2 \leq i \leq N$. This expression further simplifies to

$$\beta_i - \delta_{i+1} = v + \sum_{j=i+1}^{N} \lambda_j$$

for all $1 \leq i \leq N$ since

$$\beta_1 - \delta_2 = v + \sum_{j=2}^{N} \lambda_j.$$

Therefore we obtain that the condition

$$\beta_i - \delta_{i+1} = v + \sum_{j=i+1}^{N} \lambda_j \tag{2.15}$$

for all $1 \leq i \leq N$ ensures the validity of Danckwerts' law. After rearranging Equation (2.15) to

$$\beta_i - (\delta_{i+1} + \sum_{j=i+1}^{N} \lambda_j) = v$$

for all $1 \leq i \leq N$ we recognize the representation from Lemma 2.2.3.

2.3 Continuous Case

After presenting the discrete case we turn towards the continuous case. Physical systems are often continuous in nature. Especially time is regarded as a continuous quantity and even dividing space into separate chunks might be an assumption. This makes continuous models appealing. We present conditions for Danckwerts' law to hold in a continuous framework. Gibilaro's extension of Danckwerts' law is covered as before. To perform a continuous approach we model time by the nonnegative real numbers $[0, \infty)$ and the processing vessel by the compact interval $[0, 1]$.

From now on we consider a Markov process with continuous paths on $[0, 1]$ with a reflecting boundary at the entrance and an absorbing boundary at the exit. Transport phenomena in a vessel involving some continuous and steady throughflow are usually modelled via a diffusion process in the continuous case. Therefore we consider such a diffusion process in the following.

The theory of strongly continuous semigroups of linear operators on Banach spaces is used to establish the desired result after the connection between Markov processes and strongly continuous semigroups has been elucidated. This requires a certain understanding of diffusion processes which is discussed in the beginning of the section.

2.3.1 Diffusions

A diffusion or diffusion process $(X_t)_{t \geq 0}$ is a Markov process with continuous sample paths. There are several different ways to introduce diffusion processes, all leading essentially to the same result.

Itô's Definition as Solution to a Stochastic Differential Equation

A diffusion process can be regarded as a process that behaves locally like a Brownian Motion. Therefore it could be characterized as the solution of the stochastic differential equation

$$dX_t = v(X_t)dt + \sqrt{D(X_t)}dB_t$$

with given functions v, D and standard Brownian Motion $(B_t)_{t\geq 0}$ in the sense of K. Itô (1944). By the Itô-calculus continuous paths for this solution are available for some version of $(X_t)_{t\geq 0}$. Given the history $\{X_s : 0 \leq s \leq t\}$ of X_s until time t the displacement $dX_t := X_{t+dt} - X_t$ in the interval $(t, t+dt]$ is $v(X_t)dt + \sqrt{D(X_t)}(B_{t+dt} - B_t)$, a Gaussian random variable with expected value $v(X_t)dt$ and variance $D(X_t)dt$. We note that a Gaussian random variable is completely determined by its first two moments, i.e. here v and D and that D should be positive as the variance of a random variable.

Feller's Definition in Terms of Truncated Moments

Another possible approach to characterize diffusion processes would be to demand conditions on the truncated conditional moments of the increments of a Markov process with continuous paths in the following form

$$E[(X_{t+s} - X_s)\mathbb{1}_{\{|X_{t+s}-X_s|\leq \varepsilon\}}|X_s = x] = tv(x) + o(t),$$

$$E[(X_{t+s} - X_s)^2\mathbb{1}_{\{|X_{t+s}-X_s|\leq \varepsilon\}}|X_s = x] = tD(x) + o(t),$$

$$P(|X_{t+s} - X_s| > \varepsilon|X_s = x) = o(t),$$

with functions v, D as $t \downarrow 0$ and $\varepsilon > 0$. We see that these conditions are consistent with the above characterization via the Itô-calculus. The solution of the stochastic differential equation above satisfies these moment conditions here and the determining first two moments are considered. This characterization of diffusion processes can e.g. be found in Feller (1971).

Definition as Limit of Discrete Markov Processes

Diffusion processes can also be understood as limits of Markov birth-death chains. Consider a Markov chain $(\tilde{X}_n^\Delta)_{n\geq 0}$ on some space $\{i\Delta : 0 \leq i \leq N\}$ for $\Delta = \frac{1}{N}$ with transition probabilities

$$p_{i,i-1} := \delta_i \qquad := \tfrac{\varepsilon}{2\Delta^2}D(i\Delta) - \tfrac{\varepsilon}{2\Delta}v(i\Delta),$$

$$p_{i,i+1} := \beta_i \qquad := \tfrac{\varepsilon}{2\Delta^2}D(i\Delta) + \tfrac{\varepsilon}{2\Delta}v(i\Delta),$$

$$p_{i,i-1} := 1 - \delta_i - \beta_i \;\; = \;\; 1 - \tfrac{\varepsilon}{\Delta^2}D(i\Delta)$$

for some functions v, D and appropiate boundary conditions. We note that the conditional expected values for the one-step-displacement and its square are given as

$$E[\tilde{X}_n^\Delta - \tilde{X}_{n-1}^\Delta | \tilde{X}_{n-1}^\Delta = i\Delta] \;\; = \;\; \Delta(\beta_i - \delta_i) \;\; = \;\; \varepsilon v(i\Delta),$$

$$E[(\tilde{X}_n^\Delta - \tilde{X}_{n-1}^\Delta)^2 | \tilde{X}_{n-1}^\Delta = i\Delta] \;\; = \;\; \Delta^2(\beta_i + \delta_i) \;\; = \;\; \varepsilon D(i\Delta)$$

which corresponds to all other characterizations above. The mean displacement and the mean squared displacement of the process in one timestep are both equal to the length ε of the timestep times some function v or D respectively such that these functions determine the first two moments of the process. Now setting $\varepsilon := \tfrac{\Delta^2}{\sup D}$ and defining $X_t^\Delta := \tilde{X}_{[t/\varepsilon]}^\Delta$ leads to a limit process $X_t := w - \lim_{N \to \infty} X_t^\Delta$ ($t \geq 0$) in continuous time and space. This is also a candidate for a diffusion process and presents a bridge to Section 2.2. Details and proofs of convergence can be found in Bhattacharya and Waymire (1990) and Stroock and Varadhan (1979).

2.3.2 Diffusions and Semigroup Theory

At last we give another characterization of a diffusion process which will turn out to be the one used further and provide a link to semigroup theory. A Markov process $(X_t)_{t \geq 0}$ on some space is completely described by its starting distribution and its transition probability distributions $(p_t)_{t \geq 0}$, a Markov transition function from Section 1.1. Thus we consider the following definition.

Definition 2.3.1 *Given some Markov process $(X_t)_{t \geq 0}$ with transition probability distributions $(p_t(x, dy))_{t \geq 0}$ define the set of linear operators $(T(t))_{t \geq 0}$ via*

$$(T(t)f)(x) := E[f(X_t)|X_0 = x] = \int f(y) p_t(x, dy)$$

for $f \in L^1$ and $t \geq 0$.

We recognize two quite obvious but important facts and summarize these in Remark 2.3.2.

Remark 2.3.2 *Consider the same situation as in Definition 2.3.1 above.*

1. *The probability that the process is located in a set B at time t when starting at x at time 0 is obtained by plugging the indicator function of B into the operator T(t) above, i.e.*

$$P(X_t \in B | X_0 = x) = \int \mathbb{1}_B(y) p_t(x, dy) = (T(t) \mathbb{1}_B)(x).$$

2. *The Chapman-Kolmogorov equation $p_{t+s}(x, B) = \int p_s(y, B) p_t(x, dy)$, i.e. the Markov property, implies the semigroup functional equation*

$$T(t+s) = T(t)T(s)$$

for the set $(T(t))_{t \geq 0}$.

Thus we obtain a semigroup of linear operators $(T(t))_{t \geq 0}$ which is well-defined on L^1. To take advantage of this semigroup attribute we need some continuity property. The proper concept in this context is strong continuity.

Definition 2.3.3 *A semigroup $(T(t))_{t \geq 0}$ of linear operators on some Banach space $(X, \|\cdot\|)$ such that*

$$\lim_{t \downarrow 0} \| T(t)f - f \| = 0$$

holds for all $f \in X$ is called strongly continuous.

Checking strong continuity with $f = \mathbb{1}_{\{x\}}$ at x on the above semigroup yields

$$\lim_{t \downarrow 0} |(T(t) \mathbb{1}_{\{x\}})(x) - \mathbb{1}_{\{x\}}(x)| = \lim_{t \downarrow 0} |p_t(x, \{x\}) - 1| \stackrel{!}{=} 0 \qquad (2.16)$$

when using e.g. the sup-norm. Since even the Brownian motion does not satisfy Equation (2.16) we narrow the space of allowed functions for $(T(t))_{t \geq 0}$ down to spaces of bounded and continuous functions. The functional semigroup equation indicates that the behavior of the semigroup of linear operators $(T(t))_{t \geq 0}$ near $t = 0$ should completely determine the semigroup and leads to the notion of the generator.

Definition 2.3.4 *Let $(T(t))_{t\geq 0}$ be a strongly continuous semigroup on some Banach space X. The linear operator $(A, D(A))$ defined on $D(A) := \{f \in X : \lim_{t\downarrow 0} \frac{T(t)f-f}{t} \text{ exists}\}$ by*

$$Af := \lim_{t\downarrow 0} \frac{T(t)f - f}{t}$$

for all $f \in D(A)$ is called the generator of the semigroup $(T(t))_{t\geq 0}$.

It is a known fact in the theory of semigroups that there exists a one-to-one correspondence between strongly continuous semigroups $(T(t))_{t\geq 0}$ and their generators $(A, D(A))$, see e.g. Engel and Nagel (2000) Theorem II.1.4. Now suppose moment conditions as above and use the Taylor expansion of $f(X_t)$ around x to obtain

$$\begin{aligned}
(Af)(x) &= \lim_{t\downarrow 0} \frac{(T(t)f)(x) - f(x)}{t} \\
&= \lim_{t\downarrow 0} \frac{E[f(X_t) - f(x)|X_0 = x]}{t} \\
&= \lim_{t\downarrow 0} \frac{E[f(x) + (X_t - x)f'(x) + \frac{1}{2}(X_t - x)^2 f''(x) + R_3(x) - f(x)|X_0 = x]}{t} \\
&= v(x)f'(x) + \frac{1}{2}D(x)f''(x)
\end{aligned}$$

with functions v, D as above and some remainder term $R_3(x) = o(t)$ for suitable f. Thus there is a connection between a generator A, some semigroup $(T(t))_{t\geq 0}$ and a Markov process $(X_t)_{t\geq 0}$ belonging to some Markov transition function $(p_t)_{t\geq 0}$ as sketched in Figure 2.6. We have seen how to construct a semigroup $(T(t))_{t\geq 0}$ out of a Markov process $(X_t)_{t\geq 0}$ and ask: Does there exist a Markov process $(X_t)_{t\geq 0}$ on some space S belonging to a given semigroup $(T(t))_{t\geq 0}$ which further belongs to a generator A on some $D(A) \subset C(S)$ of the form

$$Af = vf' + \frac{1}{2}Df'' \tag{2.17}$$

as above? Fortunately the answer is yes (at least under quite general and reasonable assumptions). This can be made rigorous as in Lamperti (1977), see e.g. Theorems 1 and 2 in Chapter 7.7.

Consequently a possible definition for a diffusion process centers around the Kolmogorov backward equation (2.18).

$$(Af)(x) = \lim_{t\downarrow 0} \tfrac{(T(t)f)(x)-f(x)}{t} \qquad (T(t)f)(x) = \int f(y)p_t(x,dy)$$

$$A \longleftarrow (T(t))_{t\geq 0} \longleftarrow (X_t)_{t\geq 0}$$

Figure 2.6: Connection between generators, strongly continuous semigroups and Markov processes.

Diffusion as Solution of the Kolmogorov Backward Equation

Let $(X_t)_{t\geq 0}$ be a Markov process with continuous sample paths on some interval $S \subset \mathbb{R}$ and transition probability distributions $p_t(x,dy)$ such that $u(t,x) := \int f(y)p_t(x,dy)$ solves

$$\frac{\partial}{\partial t}u(t,x) = \frac{1}{2}D(x)\frac{\partial^2}{\partial x^2}u(t,x) + v(x)\frac{\partial}{\partial x}u(t,x) \tag{2.18}$$

with boundary condition $u(0,x) = f(x)$ for all f in a suitable subspace of $C(S)$ and given functions v and D. Then $(X_t)_{t\geq 0}$ is called a diffusion.

A different formulation of the characterization above is given below.

Remark 2.3.5 *Another way of writing Equation (2.18) in terms of semigroups is*

$$\begin{aligned}
\tfrac{\partial}{\partial t}(T(t)f)(x) &= \tfrac{\partial}{\partial t}u(t,x) \\
&= \tfrac{1}{2}D(x)\tfrac{\partial^2}{\partial x^2}u(t,x) + v(x)\tfrac{\partial}{\partial x}u(t,x) \\
&= (Au)(t,x) \\
&= (A(T(t)f))(x)
\end{aligned}$$

with $(T(t))_{t\geq 0}$ and A as in Definition 2.3.1 and Equation (2.17).

These equations above are known as the Kolmogorov backward equation. Now we like to present an example.

Example 2.3.6 *Let us look on standard Brownian motion $(B_t)_{t\geq 0}$ on $\mathbb{R}$. It obviously is a solution of the stochastic differential equation*

$$dB_t = v(X_t)dt + \sqrt{D(X_t)}dB_t$$

for $v \equiv 0$ and $D \equiv 1$. It has continuous sample paths, is Markovian and satisfies

$$E[B_{t+s} - B_s | B_s = x] = 0,$$
$$E[(B_{t+s} - B_s)^2 | B_s = x] = t,$$
$$P(|B_{t+s} - B_s| > \varepsilon | B_s = x) = 1 - \int_{-\varepsilon}^{\varepsilon} \frac{1}{\sqrt{2\pi t}} e^{-\frac{y^2}{2t}} dy = o(t).$$

The transition probability distributions of $(B_t)_{t \geq 0}$ are given by $p_t(x, dy) = \frac{1}{\sqrt{2\pi t}} e^{-\frac{(y-x)^2}{2t}} dy$. Therefore the function $u(t, x) := \int_{-\infty}^{\infty} f(y) \frac{1}{\sqrt{2\pi t}} e^{-\frac{(y-x)^2}{2t}} dy$ solves the Kolmogorov backward equation

$$\frac{\partial}{\partial t} u(t, x) = \frac{1}{2} \frac{\partial^2}{\partial x^2} u(t, x)$$

with boundary condition $u(0, x) = f(x)$ for all $f \in C_b^2(\mathbb{R})$.

2.3.3 Main Result and Further Structure of this Chapter

As in Section 2.2 we are able to state and prove an extension of Danckwerts' law.

Theorem 2.3 *For the diffusion process $(X_t)_{t \geq 0}$ on $[0, 1]$ associated to Equation (2.18) with Conditions (2.38) as in Section 2.3.5 and all $0 \leq a < b \leq 1$ the following holds*

$$E\left[\lambda^1(\{t : X_t \in [a, b]\}) \mid X_0 = 0\right] = \frac{b - a}{v_1}.$$

Here λ^1 denotes the one-dimensional Lebesgue measure and the constant $v_1 \in (0, \infty)$ describes the in- and outflow. For $a = 0$ and $b = 1$ we obtain Dankwerts' law.

The above theorem is a continuous analogue to Theorem 2.1 in Section 2.2. For the proof some effort has to be made which can be split in different parts and has value in its own right. The following list gives an outline.

1. We start by stating Kolmogorov's backward equation with the corresponding absorbing and reflecting boundary conditions. This sets the model in the right background of diffusion processes and provides the tools of semigroup theory.

2. We introduce conditions which ensure that Danckwerts' law holds and (hopefully) convince the reader of their reasonability when shortly discussing these. A derivation starting with a discrete Markov chain motivates the conditions and links this section to Section 2.2.

3. We continue by translating the situation into semigroup theory language and making tools from semigroup and operator theory available. A classification of the boundaries is presented. Afterwards we discover the semigroup which solves Kolmogorov's backward equation.

4. With this semigroup and especially after calculating the inverse of the generator of this semigroup it is a small step to prove the announced extension of Danckwerts' law.

2.3.4 Kolmogorov's Backward Equation with Boundary Conditions

As stated above we consider the Kolmogorov backward equation of diffusion

$$\frac{\partial}{\partial t}u(t,x) = \frac{1}{2}D(x)\frac{\partial^2}{\partial x^2}u(t,x) + v(x)\frac{\partial}{\partial x}u(t,x)$$

on $[0,1]$ with a reflecting boundary condition at 0 and an absorbing boundary condition at 1, i.e.

$$\begin{aligned}\frac{\partial}{\partial x}u(t,x)|_{x=0} &= 0, \\ \lim_{x \nearrow 1} u(t,x) &= 0.\end{aligned} \quad (2.19)$$

Boundary Conditions by Approximation from the Discrete Case

We like to attach some motivation for the choice of the boundary conditions above. Because the following shall only serve as a motivation we do not go into all the details concerning the convergence and existence of the discussed quantities.

Consider the approach from Section 2.3.1 to characterize a diffusion by the limit of a Markov birth-death chain. The interval $[0,1]$ might be discretized in N parts of length $\Delta = \frac{1}{N}$ forming a state space $\{0, \frac{1}{N}, \frac{2}{N}, \ldots, 1\}$ isomorph to $S = \{0, 1, 2, \ldots, N\}$.

A reflecting boundary at the point 0 is given if and only if the transition probabilities $p_{ij}(\Delta)$ for the Markov birth-death chain $(X_n^\Delta)_{n \in \mathbb{N}}$ on S suffice

$$p_{0,0}(\Delta) = 1 - p_{0,1}(\Delta) \tag{2.20}$$

with $p_{0,1}(\Delta) > 0$. We indicate the fact that the Markov birth-death chain depends on the discretization fineness by adding Δ in the notation of all relevant quantities. Now denote the n-times transition probabilities by $p_{ij}^{(n)}(\Delta) = P(X_n^\Delta = j | X_0^\Delta = i)$ and derive at the boundary

$$p_{0i}^{(n+1)}(\Delta) = \sum_{k=0}^{N} p_{0k}(\Delta) p_{ki}^{(n)}(\Delta)$$

with the Chapman-Kolmogorov equation. Since $(X_n^\Delta)_{n \in \mathbb{N}}$ is a birth-death chain we obtain

$$\begin{aligned} p_{0i}^{(n+1)}(\Delta) &= p_{00}(\Delta) p_{0i}^{(n)}(\Delta) + p_{01}(\Delta) p_{1i}^{(n)}(\Delta) \\ &= (1 - p_{01}(\Delta)) p_{0i}^{(n)}(\Delta) + p_{01}(\Delta) p_{1i}^{(n)}(\Delta) \end{aligned}$$

for all $1 \leq i \leq N$ via Equation (2.20). Inserting the definition of $p_{01}(\Delta)$ as in Section 2.3.1 and rearranging yields

$$\begin{aligned} p_{0i}^{(n+1)}(\Delta) - p_{0i}^{(n)}(\Delta) &= p_{01}(\Delta) p_{1i}^{(n)}(\Delta) - p_{01}(\Delta) p_{0i}^{(n)}(\Delta) \\ &= \beta_0 \left(p_{1i}^{(n)}(\Delta) - p_{0i}^{(n)}(\Delta) \right) \\ &= \left(\frac{\varepsilon}{2\Delta^2} D(0) + \frac{\varepsilon}{2\Delta} v(0) \right) \left(p_{1i}^{(n)}(\Delta) - p_{0i}^{(n)}(\Delta) \right) \end{aligned} \tag{2.21}$$

where we assumed $\sup_{x \in [0,1]} D(x) = 1$ to keep the presentation simple. We introduce the approximate density $p(n\varepsilon; i\Delta, j\Delta)$ by spreading the probability $p_{ij}^{(n)}(\Delta)$ over one interval of length Δ as

$$p(n\varepsilon; i\Delta, j\Delta) = \frac{p_{ij}^{(n)}(\Delta)}{\Delta} \tag{2.22}$$

for all $i, j \in S$ and $n \geq 0$. Dividing Equation (2.21) by ε and inserting the approximate density as in Equation (2.22) we get

$$\Delta \frac{p((n+1)\varepsilon; 0, i\Delta) - p(n\varepsilon; 0, i\Delta)}{\varepsilon} = \frac{1}{2} D(0) \frac{p(n\varepsilon; \Delta, i\Delta) - p(n\varepsilon; 0, i\Delta)}{\Delta} \\ + \frac{1}{2} v(0) \left(p(n\varepsilon; \Delta, i\Delta) - p(n\varepsilon; 0, i\Delta) \right)$$
(2.23)

for all $1 \leq i \leq N$ which tends to

$$0 = \frac{1}{2} D(0) \left. \frac{\partial}{\partial x} p_t(x, y) \right|_{x=0}$$
(2.24)

for all $y \in [0, 1]$ under certain assumptions as $N \to \infty$. Here $p_t(x, y)$ denotes the density of the absolutely continuous Markov transition function $(p_t)_{t \geq 0}$, i.e.

$$p_t(x, A) = \int_A p_t(x, y) dy$$

holds for all $x \in S$, $t \geq 0$ and measurable sets A. We recognize that Equation (2.24) is an analogue to the first Equation in (2.19) if $D(0) \neq 0$. In the case $D(0) = 0$ Equation (2.23) provides no boundary condition. In this case the boundary point $x = 0$ is not accessible, i.e. the probability that it is reached in finite time is zero. Therefore it is not necessary to impose a boundary condition to determine the process. In Section 2.3.6 some discussion of this matter is presented.

To examine the other boundary we keep the notation and setting used above. The same arguments as above yield

$$p_{N-1,N}^{(n+1)}(\Delta) - p_{N-1,N}^{(n)}(\Delta) = \beta_{N-1} p_{N,N}^{(n)}(\Delta) + \delta_{N-1} p_{N-2,N}^{(n)}(\Delta) - (\delta_{N-1} + \beta_{N-1}) p_{N-1,N}^{(n)}(\Delta) \\ = \beta_{N-1} \left(1 - p_{N-1,N}^{(n)}(\Delta)\right) + \delta_{N-1} \left(p_{N-2,N}^{(n)}(\Delta) - p_{N-1,N}^{(n)}(\Delta)\right)$$

since the state N is absorbing. After multiplying by $\frac{\Delta}{\varepsilon}$, inserting the definition for the transition parameters and applying Equation (2.22) this leads further to

$$\Delta^2 \frac{p((n+1)\varepsilon; (N-1)\Delta, N\Delta) - p(n\varepsilon; (N-1)\Delta, N\Delta)}{\varepsilon}$$

$$= \frac{1}{2} D((N-1)\Delta) \left(1 - p(n\varepsilon; (N-1)\Delta, N\Delta)\right) \\ + \frac{\Delta}{2} v((N-1)\Delta) \left(1 - p(n\varepsilon; (N-1)\Delta, N\Delta)\right) \\ + \frac{1}{2} D((N-1)\Delta) \left(p(n\varepsilon; (N-2)\Delta, N\Delta) - p(n\varepsilon; (N-1)\Delta, N\Delta)\right) \\ + \frac{\Delta}{2} v((N-1)\Delta) \left(p(n\varepsilon; (N-2)\Delta, N\Delta) - p(n\varepsilon; (N-1)\Delta, N\Delta)\right)$$
(2.25)

which tends to
$$0 = \lim_{x \nearrow 1} \left(\frac{1}{2} D(x) \left(1 - p_t(x, 1)\right)\right) \tag{2.26}$$

with $p_t(x, 1)$ the probability density of $p_t(x, \cdot)$ under certain assumptions as above while $N \to \infty$. We detect Equation (2.26) to be an analogue to the second Equation in (2.19).

For the case of $D(1) = v(1) = 0$ and $\lim_{x \nearrow 1} \frac{D(x)}{1-x} = -D'(1) \in (0, \infty)$ we remark that Equation (2.25) still leads to essentially the same boundary condition since dividing Equation (2.25) by Δ provides the result

$$0 = \lim_{x \nearrow 1} \left(\frac{1}{2} (D'(x) \left(1 - p_t(x, 1)\right)\right)$$

which is also an analogue to the second Equation in (2.19) if $\lim_{x \nearrow 1} D'(x) \neq 0$.

2.3.5 Conditions

This section is divided into two parts. In the first part 'Deriving the conditions' the conditions for Danckwerts' law to hold are derived based on observations in the discrete case. This part should mainly serve as motivation for choosing the conditions later on as it is done. Special attention is given to situations where the diffusion becomes degenerate at the boundaries, i.e. $D(x) = 0$ for $x = 0$ or $x = 1$. These situations determine the nature of the boundaries as discussed later in Section 2.3.6. In the second part 'Conditions' the derived conditions are collected in the form they are used in Theorem 2.3. In opposite to the first part the second part of this section exhibits precise notions whereas the first part gets along with being a more or less heuristic approach. Certainly this heuristic approach could be made precise as well.

Deriving the Conditions

We remember the conditions for Danckwerts' law to hold from Sections 2.2.1 and 2.2.3 as

(a) $\sum_{j=0}^{N} p_{j0} = 1 - v$,

(b) $\sum_{j=0}^{N} p_{ji} = 1$ for all $1 \leq i \leq N$

for a Markov chain $(X_n)_{n\geq 0}$ on $S = \{0, 1, 2, \ldots, N+1\}$ with in-/outflow $v \in (0, 1]$ and an absorbing exit at $N+1$. In Section 2.2.3 it is shown that the conditions (a) and (b) above imply a third condition

(c) $\sum_{j=0}^{N} p_{jN+1} = v$

for the flow into the exit. Let us start by considering condition (b). Applying the setup from Section 2.3.1 for the approximation of a diffusion via Markov birth-death chain we calculate

$$\sum_{j=0}^{N} p_{ji} = 1$$
$$\Leftrightarrow p_{i-1\,i} + p_{ii} + p_{i\,i+1} = 1$$
$$\Leftrightarrow \beta_{i-1} + (1 - \beta_i - \delta_i) + \delta_{i+1} = 1$$
$$\Leftrightarrow \beta_{i-1} - \beta_i - \delta_i + \delta_{i+1} = 0$$

for all $1 \leq i \leq N$. Inserting the definitions for the transition probabilities we obtain

$$\begin{aligned} 0 &= \tfrac{\varepsilon}{2\Delta^2} D((i-1)\Delta) + \tfrac{\varepsilon}{2\Delta} v((i-1)\Delta) - \tfrac{\varepsilon}{2\Delta^2} D(i\Delta) \\ &\quad - \tfrac{\varepsilon}{2\Delta^2} D(i\Delta) + \tfrac{\varepsilon}{2\Delta^2} D((i+1)\Delta) - \tfrac{\varepsilon}{2\Delta} v((i+1)\Delta) \\ &= \varepsilon \left[\tfrac{1}{2} \tfrac{D((i+1)\Delta) - 2D(i\Delta) + D((i-1)\Delta)}{\Delta^2} - \tfrac{v((i+1)\Delta) - v((i-1)\Delta)}{2\Delta} \right] \end{aligned} \quad (2.27)$$

for all $1 \leq i \leq N$. After dividing Equation (2.27) by ε it converges to

$$0 = \frac{1}{2} \frac{d^2}{dx^2} D(x) - \frac{d}{dx} v(x) \quad (2.28)$$

for $\Delta \to 0$ and all $x \in (0, 1)$ under certain assumptions.

At the boundaries the in- and outflow v is constant for the Markov chain $(X_n)_{n\geq 0}$. While investigating convergence we consider a sequence of Markov chains $(\tilde{X}_n^\Delta)_{n\geq 0}$ with space discretization Δ and time discretization $\varepsilon \sim \Delta^2$. Therefore the in- and outflow has to be adjusted $v \rightsquigarrow \frac{\varepsilon}{\Delta}v$ to take this into account. We remark that in this situation v needs not to suffice $v \in (0,1]$ but $v \in (0,\infty)$. Condition (a) yields

$$\sum_{j=0}^{N} p_{j0} = 1 - \tfrac{\varepsilon}{\Delta}v$$
$$\Leftrightarrow p_{00} + p_{10} = 1 - \tfrac{\varepsilon}{\Delta}v$$
$$\Leftrightarrow 1 - \beta_0 + \delta_1 = 1 - \tfrac{\varepsilon}{\Delta}v$$
$$\Leftrightarrow \beta_0 - \delta_1 = \tfrac{\varepsilon}{\Delta}v$$

at the entrance. Again inserting the definitions for the transition probabilities and rearranging leads to

$$\begin{aligned}\tfrac{\varepsilon}{\Delta}v &= \tfrac{\varepsilon}{2\Delta^2}D(0) + \tfrac{\varepsilon}{2\Delta}v(0) - \tfrac{\varepsilon}{2\Delta^2}D(\Delta) + \tfrac{\varepsilon}{2\Delta}v(\Delta) \\ &= \tfrac{\varepsilon}{\Delta}\left[\tfrac{1}{2}\tfrac{D(0)-D(\Delta)}{\Delta} + \tfrac{1}{2}(v(0) + v(\Delta))\right].\end{aligned} \quad (2.29)$$

We divide Equation (2.29) by $\frac{\varepsilon}{\Delta}$ and get

$$v = -\frac{1}{2}\frac{d}{dx}D(x)\bigg|_{x=0} + v(0) \quad (2.30)$$

as the limit for $\Delta \to 0$ under certain assumptions. We remark that even in the case $D(0) = 0$ Equation (2.29) shows that the derivative of the diffusion function D exists near zero, i.e. $\lim_{x \searrow 0} D'(x) \in [0, \infty)$. The positivity of the limit follows from the positivity of the diffusion function D on $(0,1)$.

We operate similar to obtain with condition (c)

$$\sum_{j=0}^{N} p_{jN+1} = \tfrac{\varepsilon}{\Delta}v$$
$$\Leftrightarrow p_{NN+1} = \tfrac{\varepsilon}{\Delta}v$$
$$\Leftrightarrow \beta_{NN+1} = \tfrac{\varepsilon}{\Delta}v \quad (2.31)$$
$$\Leftrightarrow \tfrac{\varepsilon}{2\Delta^2}D(N\Delta) + \tfrac{\varepsilon}{\Delta}v(N\Delta) = \tfrac{\varepsilon}{\Delta}v$$

an equation for the exit. The expression for β_{NN+1} was slightly modified from the one given in Section 2.3.1 to level the terms of order $\frac{\varepsilon}{\Delta}$. We multiply Equation (2.31) by $\frac{\Delta^2}{\varepsilon}$

and insert the definitons for the transition probabilities to get

$$\Delta v = \frac{1}{2}D(N\Delta) + \Delta v(N\Delta) \tag{2.32}$$

which tends to

$$0 = \frac{1}{2}D(1) \tag{2.33}$$

for $\Delta \to 0$ under certain assumptions. Now Equation (2.31) becomes

$$\frac{\varepsilon}{\Delta}v = \frac{\varepsilon}{\Delta}\frac{1}{2}\frac{D(N\Delta) - D(1)}{\Delta} + \frac{\varepsilon}{\Delta}v(N\Delta) \tag{2.34}$$

because $D(1) = 0$ and tends to

$$v = -\frac{1}{2}\frac{d}{dx}D(x)\bigg|_{x=1} + v(1) \tag{2.35}$$

after multiplying with $\frac{\Delta}{\varepsilon}$ for $\Delta \to 0$ under certain assumptions. Equation (2.35) is rederived by an other observation below in Equation (2.36).

A further consequence of Equation (2.34) is the existence of the limit of the derivative of the diffusion function D near one, i.e. $\lim_{x \nearrow 1} D'(x) \in \mathbb{R}$. If $D(1) = 0$ we conclude that $D'(1) \in (-\infty, 0]$ since D is positive on $(0, 1)$.

Connecting Equation (2.28) inside with those at the boundaries of the interval $[0, 1]$ via integration we deduce

$$\begin{aligned}\tfrac{1}{2}D'(x) - v(x) &= \tfrac{1}{2}D'(0) - v(0) &= \tfrac{1}{2}D'(1) - v(1) \\ \Leftrightarrow \tfrac{1}{2}D'(x) - v(x) &= -v &= -v\end{aligned} \tag{2.36}$$

which is equivalent to

$$v(x) = \frac{1}{2}D'(x) + v \tag{2.37}$$

for all $x \in [0, 1]$ when implementing the boundary conditions (2.30) and (2.35). In many physical models it is not possible to leave the vessel via diffusion only via drift. Then the derivatives of the diffusion function at the boundaries become zero, i.e. $D'(0) = D'(1) = 0$. As a consequence the drift function v at the boundaries suffices $v(0) = v(1) = v$. This complies with our intuition about the physical process.

Conditions

Everything has now been done to formulate precisely all conditions needed for Danckwerts' law to hold. The diffusion function D should be continuously differentiable on $[0,1]$ and positive (at least in the interior) since it determines the variance of the process. At the absorbing boundary (the exit) we need some control of D and require it to attain the value zero. The drift function v should be continuous. The mass balance equation

$$v(x) = \frac{1}{2}D'(x) + v_1$$

has to hold for all $x \in [0,1]$ with the in-/outflow parameter $v_1 \in (0, \infty)$. The in-/outflow parameter v_1 describes the constant in- and outflow as the required premise in Danckwerts' law. Here we changed the notation for the in-/outflow parameter to v_1 to avoid misunderstandings.

Summing up we consider a continously differentiable diffusion function with value zero at the exit and existing derivatives at the boundaries which is positive in the interior and a continuous drift function v sufficing equation (2.37). In symbols we have

1. $D \in C^1([0,1])$,
2. $D(x) > 0$ for all $x \in (0,1)$,
3. $D(1) = 0$,   (2.38)
4. $v \in C([0,1])$,
5. $v(x) = \frac{1}{2}D'(x) + v_1$ for all $x \in [0,1]$,

with $v_1 \in (0, \infty)$ the constant describing the in- and outflow. Conditions 1. and 3. control the behavior of the diffusion at the boundaries for a degenerate diffusion function D. From now on we tacitly assume that all the conditions above hold.

2.3.6 Deriving the Semigroup

The model is set up and all needed conditions are given. Now it is time to start proving Theorem 2.3. As announced before we begin by translating the situation into the language of strongly continuous semigroups. A candidate for the generator A of the sought semigroup is easily found for which its inverse is computed explicitly. Thereafter the generator property of A is verified. Finally the semigroup itself is shown to be associated to a Markov transition function and thus to the Markov process belonging to the Kolmogorov backward equation under the presented conditions. We like to point out that therewith the existence of a solution of the Kolmogorov backward equation is proved as well. These steps prepare the proof of Theorem 2.3 in Section 2.3.7. For the following let $x_0 \in (0,1)$.

Translation into Semigroup Theory Language

Obviously the Kolmogorov backward equation (2.18) corresponds to a differential operator. Having in mind the boundary conditions from Section 2.3.5 and the fact that the generator of a strongly continuous semigroup has a dense domain we set

$$C_0([0,1]) = \{u : [0,1] \to \mathbb{R} \mid u \text{ continuous}, u(1) = 0\},$$

$$m(x) = \tfrac{1}{2}D(x) \quad \text{for all } x \in [0,1],$$

$$q(x) = v(x) = \tfrac{1}{2}D'(x) + v_1 \quad \text{for all } x \in [0,1]$$

for some $v_1 \in (0, \infty)$ and consider the linear differential operator

$$Au = mu'' + qu'$$

defined on

$$D(A) = \{u \in C_0([0,1]) \cap C^2((0,1)) : Au \in C_0([0,1]), \lim_{x \searrow 0} u'(x) = 0\}$$

$$= \{u \in C_0([0,1]) \cap C^2((0,1)) : u'' \text{ continuous in } 0, \lim_{x \searrow 0} u'(x) = 0,$$

$$\lim_{x \nearrow 1} \tfrac{1}{2}D(x)u''(x) + (\tfrac{1}{2}D'(x) + v_1)u'(x) = 0\}$$

if $D(0) \neq 0$ and

$$D(A) = \{u \in C_0([0,1]) \cap C^2((0,1)) : Au \in C_0([0,1])\}$$

$$= \{u \in C_0([0,1]) \cap C^2((0,1)) : \lim_{x \searrow 0} \tfrac{1}{2}D(x)u''(x) + (\tfrac{1}{2}D'(x) + v_1)u'(x) \text{ exists},$$

$$\lim_{x \nearrow 1} \tfrac{1}{2}D(x)u''(x) + (\tfrac{1}{2}D'(x) + v_1)u'(x) = 0\}$$

if $D(0) = 0$. We recognize that the domain $D(A)$ of the operator A changes for different values of the diffusion function D at the boundary point $x = 0$. Nonetheless we keep only one notation for the domain $D(A)$ because all our results hold in both cases. At the adequate places in proofs we comment on differences and similarities of both cases. If $D(0) = 0$ holds then no additional condition at the left boundary $x = 0$ is necessary. This is further discussed below.

As things develop we make abundant use of the auxiliary function W, the socalled Wronskian. For some arbitrary $x_0 \in (0, 1)$ it is defined by

$$\begin{aligned}
W(x) &= \exp\{-\int_{x_0}^x \tfrac{q(s)}{m(s)} ds\} \\
&= \exp\{-\int_{x_0}^x \tfrac{\frac{1}{2}D'(s)+v_1}{\frac{1}{2}D(s)} ds\} \\
&= \exp\{-\int_{x_0}^x \tfrac{D'(s)}{D(s)} ds\} \exp\{-\int_{x_0}^x \tfrac{2v_1}{D(s)} ds\} \\
&= \exp\{\ln(D(x_0)) - \ln(D(x))\} \exp\{-2v_1 \int_{x_0}^x \tfrac{1}{D(s)} ds\} \\
&= \tfrac{D(x_0)}{D(x)} \exp\{-2v_1 \int_{x_0}^x \tfrac{1}{D(s)} ds\}
\end{aligned}$$

for all $x \in (0, 1)$. The boundary behavior of the process is closely interwoven with the finiteness of the Wronskian.

In the following we show that this linear operator $(A, D(A))$ is the generator of a positive, strongly continuous semigroup on $C_0([0, 1])$ which solves Equation (2.18) and is associated to a Markov process via a Markov transition function (see Lamperti (1977) for the case $C([0, 1])$). Therefore we will extend techniques from the case $C([0, 1])$ to the case $C_0([0, 1])$. A classification of the boundaries is presented at first.

Boundary Classification

In the original articles by William Feller (1952, 1954) the boundaries of a diffusion process associated to Komogorov's backward equation are charaterized in terms of integrability of the functions W and $\frac{1}{mW}$ near the boundaries. The boundary $b \in \{0,1\}$ is called

    (i)   *regular* if and only if $W \in L^1([x_0, b])$ and $\frac{1}{mW} \in L^1(x_0, b)$,

    (ii)   *exit* if and only if $W \in L^1([x_0, b])$ and $\frac{1}{mW} \notin L^1(x_0, b)$,

    (iii)   *entrance* if and only if $W \notin L^1([x_0, b])$ and $\frac{1}{mW} \in L^1(x_0, b)$,

    (iv)   *natural* if and only if $W \notin L^1([x_0, b])$ and $\frac{1}{mW} \notin L^1(x_0, b)$,

    (v)   *accessible* if and only if it is either regular or exit,

    (vi)   *inaccessible* if and only if it is not accessible.

We see that for entrance and natural boundaries no boundary conditions are necessary since they are inaccessible, i.e. the probability that the associated Markov process reaches them in finite time equals zero.

For regular and exit boundaries boundary conditions have to be assigned. When looking back to the definitons of the domains $D(A)$ for the operator A we expect the left boundary $x = 0$ to be regular or exit if $D(0) \neq 0$ and natural or entrance otherwise while the boundary $x = 1$ should be regular or exit. Semantics further suggest to find a regular or entrance boundary on the left side and an exit boundary on the right. We refer to Proposition 2.3.9 below.

But before we start to identify the boundaries we need some auxiliary results.

Lemma 2.3.7 *Let $f : [0, 1] \to \mathbb{R}$ be continuous on $[0, 1]$ and continuously differentiable in $(0, 1)$ and $b \in \{0, 1\}$ a boundary point. Then the following holds.*

(i) $\lim_{n \to \infty}(x_n - b)f'(x_n) = 0$,   *for some sequence $x_n \to b$*

(ii)   *Suppose f is positive in the interior as well,*   i.e. $f(x) > 0$ *for all $x \in (0, 1)$.*

    (a) $f(b) = 0$ and $f'(b)$ exists $\Leftrightarrow$ $\lim_{x \to b} \frac{f(x)}{x-b}$ *exists*

    (b) *Any condition stated in (a) holds.* $\Rightarrow$ $\frac{1}{f}$ *is not integrable at the boundary point b.*

Proof. Without loss of generality assume $b = 0$. We start by proving claim (i). Distinguish two cases $(C1)$ and $(C2)$.

Case $(C1)$: f' changes sign in all intervals $(0, \varepsilon)$. Then for all $\varepsilon > 0$ the continuity of f' in $(0, 1)$ implies the existence of $x_\varepsilon \in (0, \varepsilon)$ such that $f(x_\varepsilon) = 0$ holds. Taking the sequence $\frac{1}{n}$ for ε now yields the sought-after sequence $x_{1/n}$ with

$$\lim_{n \to \infty} x_{1/n} f'\left(x_{1/n}\right) = \lim_{n \to \infty} x_{1/n} \cdot 0 = 0$$

and $x_{1/n} \to 0$ since $x_{1/n} \in (0, 1/n)$ for all $n \geq 1$.

Case $(C2)$: There exists $\varepsilon > 0$ such that f is monotone on $(0, \varepsilon)$. Without loss of generality let $f' \geq 0$ on $(0, \varepsilon)$. Suppose $\liminf_{x \to 0} x f'(x) > 0$. Then there exist $\delta_1, \delta_2 > 0$ such that

$$x f'(x) \geq \delta_1$$

for all $x \in (0, \delta_2)$. This yields further

$$f(x) - f(0) = \int_0^x f'(s) ds \geq \int_0^x \frac{\delta_1}{s} ds = \infty$$

on $(0, \delta_2)$ in contradiction to the continuity of f on $[0, 1]$ and proves claim (i).

Consider claim (ii). Suppose that the limit $\lim_{x \searrow 0} \frac{f(x)}{x}$ exists. Therefore we get

$$f(0) = \lim_{x \searrow 0} f(x) = \lim_{x \searrow 0} x \frac{f(x)}{x} = 0$$

since f is continuous. This leads subsequently to

$$f'(0) = \lim_{x \searrow 0} \frac{f(x) - f(0)}{x} = \lim_{x \searrow 0} \frac{f(x)}{x} \tag{2.39}$$

which shows the existence of the derivative of f at 0. The converse follows directly from Equation (2.39). Let us now turn towards the case

$$\lim_{x \to 0} \frac{f(x)}{x} \text{ exists} \Rightarrow \frac{1}{f} \notin L^1([0, x_0])$$

with some $x_0 \in (0, 1)$. Let $\alpha = \lim_{x \to 0} \frac{f(x)}{x}$. Then there exists $\varepsilon > 0$ such that

$$\frac{|f(x)|}{x} \leq |\alpha| + 1$$

for all $x \in (0, \varepsilon)$. Thus it follows that
$$\frac{1}{|f(x)|} \geq \frac{1}{x(|\alpha|+1)}$$
on $x \in (0, \varepsilon)$ and proves the claim since $\frac{1}{x(|\alpha|+1)}$ is not integrable near 0. □

We continue by collecting some consequences from Section 2.3.5 and useful identities for the Wronskian.

Lemma 2.3.8 *Let $f \in D(A)$. Then the following holds*

(i) $W'(x) = -\frac{q(x)}{m(x)} W(x)$,

(ii) $\int_{x_0}^{x} W(s)ds = -\frac{m(x)W(x) - m(x_0)}{v_1}$,

(iii) $\lim_{x \searrow 0} \left[-\frac{m(x)W(x)}{v_1} \int_0^x \frac{f(s)}{m(s)W(s)} ds \right] = 0$,

$\lim_{x \nearrow 1} \left[-\frac{m(x)W(x)}{v_1} \int_0^x \frac{f(s)}{m(s)W(s)} ds \right] = 0$,

(iv) $\frac{m(x)f''(x) + q(x)f'(x)}{m(x)W(x)} = \left(\frac{f'(x)}{W(x)} \right)'$

for all $x \in (0, 1)$.

Proof. (i) We differentiate
$$W'(x) = \left(\exp\{ -\int_{x_0}^{x} \frac{q(s)}{m(s)} ds \} \right)' = -\frac{q(x)}{m(x)} W(x)$$
and obtain statement (i).

(ii) The calculation
$$(m(x)W(x))' = m'(x)W(x) + m(x)W'(x) = (m'(x) - q(x))W(x) = -v_1 W(x)$$
yields
$$\int_{x_0}^{x} W(s)ds = -\frac{1}{v_1} \int_{x_0}^{x} (m(s)W(s))' ds = -\frac{m(x)W(x) - m(x_0)}{v_1}$$
which is statement (ii).

(*iii*) Consider

$$-\frac{m(x)W(x)}{v_1}\int_0^x \frac{f(s)}{m(s)W(s)}ds = -\frac{1}{v_1}\frac{D(x_0)}{2}\exp\{-2v_1\int_{x_0}^x \frac{1}{D(s)}ds\}\int_0^x \frac{2f(s)}{D(x_0)}\exp\{2v_1\int_{x_0}^s \frac{1}{D(t)}dt\}ds$$

$$= -\frac{1}{v_1}\int_0^x f(s)\exp\{-2v_1\int_s^x \frac{1}{D(t)}dt\}ds.$$

This yields

$$\left|\lim_{x\searrow 0}\left[-\frac{m(x)W(x)}{v_1}\int_0^x \frac{f(s)}{m(s)W(s)}ds\right]\right| \leq \frac{||f||_\infty}{v_1}\lim_{x\searrow 0} x = 0$$

on the left. On the right $\frac{1}{D} \notin L^1([x_0, 1])$ by Lemma 2.3.7 (*ii*) (*b*) implies for all $\varepsilon > 0$ the existence of $x_\varepsilon < y_\varepsilon$ in $(1 - \frac{\varepsilon}{2}, 1)$ such that

$$\exp\left\{-2v_1 \int_{x_\varepsilon}^{y_\varepsilon} \frac{1}{D(t)}dt\right\} < \frac{\varepsilon}{2}$$

holds. We estimate

$$\int_0^{y_\varepsilon} \exp\left\{-2v_1 \int_s^{y_\varepsilon} \frac{1}{D(t)}dt\right\}ds \leq \int_0^{x_\varepsilon} \exp\left\{-2v_1 \int_s^{y_\varepsilon} \frac{1}{D(t)}dt\right\}ds + \int_{x_\varepsilon}^{y_\varepsilon} \exp\left\{-2v_1 \int_s^{y_\varepsilon} \frac{1}{D(t)}dt\right\}ds$$

$$\leq x_\varepsilon \frac{\varepsilon}{2} + \frac{\varepsilon}{2}$$

$$< \varepsilon$$

and obtain that there exists a sequence $y_n \to 1$ such that

$$\lim_{n\to\infty} \int_0^{y_n} \exp\left\{-2v_1 \int_s^{y_n} \frac{1}{D(t)}dt\right\}ds = 0$$

holds. By the monotony and boundedness of the considered integral we get

$$\left|\lim_{x\nearrow 1}\left[-\frac{m(x)W(x)}{v_1}\int_0^x \frac{f(s)}{m(s)W(s)}ds\right]\right| \leq \frac{||f||_\infty}{v_1}\lim_{x\nearrow 1}\int_0^x \exp\left\{-2v_1 \int_s^x \frac{1}{D(t)}dt\right\}ds = 0,$$

which is statement (*iii*) on the right.

(*iv*) Straightforward calculating results in statement (*iv*), i.e.

$$\left(\frac{f'(x)}{W(x)}\right)' = \frac{f''(x)W(x) - f'(x)W'(x)}{W^2(x)} = \frac{m(x)f''(x) + q(x)f'(x)}{m(x)W(x)}.$$

□

Now we are able to classify the boundaries as announced above.

Proposition 2.3.9 *Under the conditions in Section 2.3.5 the following holds.*

(a) *The boundary point $x = 0$ is regular if and only if $D(0) \neq 0$ and entrance otherwise, i.e. if and only if $D(0) = 0$.*

(b) *The boundary point $x = 1$ is regular or exit. If $v(1) > 0$ holds then $x = 1$ is an exit.*

Proof. We remind us of the definition of the Wronskian as

$$\begin{aligned} W(x) &= \frac{D(x_0)}{D(x)} \exp\{-2v_1 \int_{x_0}^{x} \frac{1}{D(s)} ds\} \\ &= \frac{m(x_0)}{m(x)} \exp\{-v_1 \int_{x_0}^{x} \frac{1}{m(s)} ds\} \end{aligned} \qquad (2.40)$$

for all $x \in (0,1)$. Let us first consider the left boundary $x = 0$. We see that

$$\frac{1}{mW} \in L^1([0, x_0])$$

holds since $|\frac{1}{mW}| = |\frac{1}{m(x_0)} \exp\{v_1 \int_{x_0}^{x} \frac{1}{m(s)} ds\}|$ is bounded on $[0, x_0]$ independent of the value of the diffusion function D at zero. Therefore the left boundary is either regular or entrance. Lemma 2.3.8 yields

$$\int_{x}^{x_0} W(s) ds = \frac{m(x)W(x) - m(x_0)}{v_1} \qquad (2.41)$$

for all $x \in (0,1)$. Now, by Equation (2.40) we have

$$m(x)W(x) = m(x_0) \exp\left\{v_1 \int_{x}^{x_0} \frac{1}{m(s)} ds\right\}$$

and thus

$$W \in L^1([0, x_0])$$

holds if $m(0) = \frac{1}{2} D(0) \neq 0$. Hence the boundary $x = 0$ is regular if $D(0) \neq 0$. If $D(0) = 0$ Lemma 2.3.7 provides that $\frac{1}{D}$ is not integrable and with it the function mW tends to infinity for $x \to 0$. Then

$$W \notin L^1([0, x_0])$$

holds and the boundary is entrance. This also proves the converse direction for the 'if and only if' statement because regular and entrance boundaries are exclusive.

Let us now come to the other boundary $x = 1$. Here we quickly observe that
$$W \in L^1([x_0, 1])$$
holds by Equations (2.40) and (2.41) above and the positivity of m. Hence the right boundary is either regular or exit.

Consider $v(1) > 0$. The positivity of m in $(0, 1)$ implies that $m'(1) \leq 0$ and by $v(1) = m'(1) + v_1$ that $0 \leq |m'(1)| < v_1$ holds. Since $m(1) = 0$ we get
$$\lim_{x \to 1} \frac{m(x)}{1 - x} = |m'(1)| < v_1$$
and therefore there exists $\varepsilon \in (0, 1 - x_0)$ such that
$$m(x) \leq v_1(1 - x)$$
holds on $(1 - \varepsilon, 1)$. With this and Equation (2.40) we calculate
$$\begin{aligned}
\frac{1}{m(x)W(x)} &= \frac{1}{m(x_0)} \exp\left\{v_1 \int_{x_0}^{x} \frac{1}{m(s)} ds\right\} \\
&\geq \frac{1}{m(x_0)} \exp\left\{v_1 \int_{1-\varepsilon}^{x} \frac{1}{v_1(1-s)} ds\right\} \\
&= \frac{1}{m(x_0)} \exp\left\{ln(1 - \varepsilon) - ln(1 - x)\right\} \\
&= \frac{1-\varepsilon}{m(x_0)(1-x)}
\end{aligned}$$
for all $x \in (1 - \varepsilon, 1)$. We finish the proof by concluding that the right boundary is exit because
$$\frac{1}{mW} \notin L^1([x_0, 1])$$
since $\frac{1}{1-x}$ is not integrable near 1. $\square$

Calculating the Inverse of A

Using Lemma 2.3.8 we are now able to define a linear operator $G : C_0([0, 1]) \to D(A)$ and show that it is the inverse operator of A. The strength of our approach to this problem lies in the fact that we are able to give an explicit formula for the resolvent of the generator at 0, i.e. for the inverse A^{-1} of the generator A as will be seen in Section 2.3.7.

Theorem 2.4 *The linear operator* $G : C_0([0,1]) \to D(A)$

$$(Gf)(x) := -\frac{m(x)W(x)}{v_1} \int_0^x \frac{f(s)}{m(s)W(s)} ds - \frac{1}{v_1} \int_x^1 f(s) ds$$

is the inverse of $(A, D(A))$, *i.e.* $G = A^{-1}$.

Proof. By Lemma 2.3.8 (iii) $Gf(x)$ exists for all $x \in [0,1]$ and $Gf \in C_0([0,1])$ holds for all $f \in C_0([0,1])$. Again using Lemma 2.3.8 and the definition of q and m we calculate the derivatives as

$$(Gf)'(x) = W(x) \int_0^x \frac{f(s)}{m(s)W(s)} ds \tag{2.42}$$

and

$$(Gf)''(x) = -\frac{q(x)}{m(x)} W(x) \int_0^x \frac{f(s)}{m(s)W(s)} ds + \frac{f(x)}{m(x)} \tag{2.43}$$

for all $f \in C_0([0,1])$ and $x \in (0,1)$. We have to check that $Gf \in D(A)$ holds for all $f \in C_0([0,1])$. Obviously Equations (2.42) and (2.43) imply that $A(Gf) = f$ for all $f \in C_0([0,1])$, in particular $A(Gf) \in C_0([0,1])$. Only the boundary behavior of the derivative of Gf near 0 if $m(0) \neq 0$ still has to be examined. In this case

$$\lim_{x \searrow 0} (Gf)'(x) = \lim_{x \searrow 0} \left[W(x) \int_0^x \frac{f(s)}{m(s)W(s)} ds \right] = 0$$

holds since W stays bounded by Equation (2.40) and $\frac{1}{mW} \in L^1([0, x_0])$ by Proposition 2.3.9. Thus $Gf \in D(A)$.

Calculating while using Lemma 2.3.8 (iv) we obtain

$$
\begin{aligned}
(G(Af))(x) &= -\frac{m(x)W(x)}{v_1} \int_0^x \frac{m(s)f''(s) + q(s)f'(s)}{m(s)W(s)} ds - \frac{1}{v_1} \int_x^1 (m(s)f''(s) + q(s)f'(s)) ds \\
&= -\frac{m(x)W(x)}{v_1} \int_0^x \left(\frac{f'(s)}{W(s)}\right)' ds - \frac{1}{v_1} \int_x^1 (m(s)f''(s) + (m'(s) + v_1)f'(s)) ds \\
&= -\frac{m(x)W(x)}{v_1} \left[\frac{f'(x)}{W(x)} - \lim_{s \searrow 0} \frac{f'(s)}{W(s)} \right] - f(1) + f(x) \\
&\quad - \frac{1}{v_1} \left[\lim_{s \nearrow 1} m(s)f'(s) - m(x)f'(x) \right] \\
&= f(x) - f(1) + \frac{m(x)W(x)}{v_1} \lim_{s \searrow 0} \frac{f'(s)}{W(s)} - \lim_{s \nearrow 1} \frac{m(s)f'(s)}{v_1} \\
&= f(x)
\end{aligned}
$$

for all $f \in D(A)$ and $x \in [0,1]$.

For the last step we consider the following. $f(1) = 0$ holds since $f \in C_0([0,1])$. Then we note that all considered limits exist since G is well-defined on $C_0([0,1])$. If $m(0) \neq 0$ we get

$$\lim_{s \searrow 0} \frac{f'(s)}{W(s)} = \frac{1}{W(0)} \lim_{s \searrow 0} f'(s) = 0$$

by the boundary condition for f' and the boundedness of $\frac{1}{W(0)}$ from Equation (2.40). If $m(0) = 0$ an expansion gives

$$\lim_{n \to \infty} \frac{f'(s_n)}{W(s_n)} = \lim_{n \to \infty} s_n f'(s_n) \frac{m(s_n)}{s_n} \frac{1}{m(s_n)W(s_n)} = 0$$

for some sequence $s_n \to 0$ by Lemma 2.3.7 (i) since $m'(0)$ exists and $\frac{1}{mW}$ is bounded by Equation (2.40). The existence of the limit yields that it is sufficient to consider the distinct sequence $s_n \to 0$ to conclude

$$\lim_{s \searrow 0} \frac{f'(s)}{W(s)} = 0.$$

The same reasoning applies again Lemma 2.3.7 (i), the identity $m(1) = 0$ and the existence of $m'(1)$ to find some sequence $r_n \to 1$ such that

$$\lim_{n \to \infty} \frac{m(r_n)f'(r_n)}{v_1} = \frac{1}{v_1} \lim_{n \to \infty} \frac{m(r_n) - m(1)}{r_n - 1} f'(r_n)(r_n - 1) = 0$$

holds and implies

$$\lim_{s \nearrow 1} \frac{m(s)f'(s)}{v_1} = 0.$$

Thus $G(Af) = f$ and hence $G = A^{-1}$. □

Proving the Generator Property of $(A, D(A))$

To show that $(A, D(A))$ is the generator of a positive, strongly continuous semigroup we will use the Lumer-Phillips Theorem, see e.g. Engel-Nagel (2000) Proposition II.3.14 and Theorem II.3.15.

Theorem (Lumer-Phillips) *Let $(A, D(A))$ be a densely defined, dissipative operator on X such that $\lambda - A$ is surjective for some $\lambda > 0$. Then $(A, D(A))$ generates a strongly continuous contraction semigroup on X.*

There are different equivalent conditions to define dissipativity of a linear operator. The following is used in this paper.

A linear operator $(A, D(A))$ on a Banach space X is *dissipative* if and only if for every $f \in D(A)$ there exists $j(f) \in \mathcal{J}(f) = \{f^* \in X^* :\,<f, f^*> = ||f||^2 = ||f^*||^2\}$ such that

$$Re <Af, j(f)> \leq 0$$

holds. The set $\mathcal{J}(f)$ is called the *duality set* of f in the dual space X^* of X.

The following theorem is the center of this subsection.

Theorem 2.5 *The linear operator $(A, D(A))$ is the generator of a positive, strongly continuous semigroup on $C_0([0, 1])$.*

Proof. It is known that $C_0^\infty([0, 1])$ is dense in $C_0([0, 1])$. Clearly $C_0^\infty([0, 1]) \subset D(A)$ holds as well. Hence $D(A)$ is dense in $C_0([0, 1])$.

To verify the surjectivity of $\lambda - A$ for some $\lambda > 0$ we use the existence of A^{-1}. Since $G = A^{-1}$ is bounded there exists $\lambda > 0$ such that $\lambda ||A^{-1}|| < 1$ and therefore $(Id - \lambda A^{-1})^{-1}$ exists which implies the existence of $(\lambda - A)^{-1} = -A^{-1}(Id - \lambda A^{-1})^{-1}$.

Finally dissipativity of A needs to be shown. Let $f \in D(A)$ and denote by

$$\mathcal{J}(f) = \{f^* \in C_0^*([0, 1]) : \langle f, f^* \rangle = ||f||_\infty^2 = ||f^*||_{C_0^*([0,1])}^2\}$$

the duality set of f in the dual space $C_0^*([0, 1])$ of $C_0([0, 1])$. For $z \in [0, 1]$ such that $|f(z)| = ||f||_\infty$ define $f_z^* : C_0([0, 1]) \to \mathbb{R}$ by $\langle f_z^*, g \rangle := f(z)g(z)$ and obtain that $f_z^* \in \mathcal{J}(f)$ since

$$\langle f_z^*, f \rangle = f(z)^2 = ||f||_\infty^2 = ||f_z^*||_{C_0^*([0,1])}^2 .$$

If $z \in (0,1)$ then $f'(z) = 0$ and $f''(z) \leq 0$ if $f(z) > 0$ or $f'(z) = 0$ and $f''(z) \geq 0$ if $f(z) < 0$. Since $m(z) \geq 0$ it holds that

$$\operatorname{Re}\langle f_z^*, Af \rangle = f(z)(Af)(z) = f(z)m(z)f''(z) \leq 0.$$

If $z = 1$ then $f(1) = 0$ implies $f \equiv 0$ and $\operatorname{Re}\langle f_1^*, Af \rangle = f(1)(Af)(1) = 0$. We note that $(Af)(1)$ exists since $Af \in C_0([0,1])$ for all $f \in D(A)$.

If $z = 0$ we distinguish between $m(0) = 0$ and $m(0) \neq 0$. If $m(0) \neq 0$ then the boundary condition $\lim_{x \searrow 0} f'(x) = 0$ is valid and f'' is continuous in 0. Hence we are in the same situation as if $z \in (0,1)$.

Otherwise $m(0) = 0$ applies. Assume that $f(0) > 0$ and $(Af)(0) > 0$. Since $Af \in C_0([0,1])$ there exists $0 < \delta < 1$ such that $0 < (Af)(x) < \infty$ for all $x \in [0, \delta)$. Thus we can integrate Af and obtain

$$\begin{aligned}
0 < \int_0^\varepsilon (Af)(x)dx &= \int_0^\varepsilon (mf'' + (m' + v_1)f')(x)dx \\
&= \int_0^\varepsilon (mf')'(x) + v_1 f'(x)dx \\
&= (mf')(\varepsilon) + v_1 f(\varepsilon) - \lim_{s \searrow 0}(mf')(s) - v_1 f(0) \\
&= v_1(f(\varepsilon) - f(0)) + (mf')(\varepsilon) - \lim_{s \searrow 0}(mf')(s)
\end{aligned} \quad (2.44)$$

for all $0 < \varepsilon \leq \delta$. In particular the limit $\lim_{s \searrow 0}(mf')(s)$ has to exist. Since f has a positive maximum at 0 there exists $0 < \delta_1 < \delta$ such that

$$0 > f(x) - f(0) = \int_0^x f'(s)ds$$

holds for all $0 < x < \delta_1$. Hence we have that $v_1(f(x) - f(0)) < 0$ holds for all $0 < x < \delta_1$ because $v_1 > 0$ and that there exists some sequence $x_n \to 0$ such that $f'(x_n) < 0$ holds for all $n \in \mathbb{N}$. By the positivity of m in $(0,1)$ it follows that $(mf')(x_n) < 0$ holds for all $n \in \mathbb{N}$ and therefore $\lim_{s \searrow 0}(mf')(s) \leq 0$. We show that $\lim_{s \searrow 0}(mf')(s) = 0$ is valid since otherwise we get a contradiction to the boundedness of f as follows.

Assume $\lim_{s \searrow 0}(mf')(s) = \alpha < 0$ holds. Using that $(mf')(s) < \frac{\alpha}{2}$ holds for all $0 < s < \delta_2$ and some $0 < \delta_2 < \delta_1$ we obtain by integration

$$f(x) - f(0) = \int_0^x f'(s)ds \leq \int_0^x \frac{\alpha}{2m(s)}ds = -\infty$$

for all $0 < x < \delta_2$ since $\frac{1}{m} \notin L_1([0, x_0])$ by Lemma 2.3.7 $(ii)(b)$. This is the announced contradiction. Altogether we have shown that

$$v_1(f(\varepsilon) - f(0)) + (mf')(\varepsilon) - \lim_{s \searrow 0}(mf')(s) < 0$$

holds for all $0 < \varepsilon \leq \delta_2$. This is a contradiction to Equation (2.44) and proves

$$Re\langle f_0^*, Af \rangle = f(0)(Af)(0) = f(0)(m(0)f''(0) + v(0)f'(0)) \leq 0$$

if $f(0) > 0$. An analoguous argument is valid for the case $f(0) < 0$. The case $f(0) = 0$ has already been handled above while the right boundary $z = 1$ has been under consideration.

Thus the operator $(A, D(A))$ is dissipative by Engel-Nagel (2000) Proposition II.3.23. This yields that $(A, D(A))$ is the generator of a strongly continuous, contraction semigroup $(T(t))_{t \geq 0}$ on $C_0([0, 1])$. $\square$

Deriving the Semigroup and its Associated Diffusion

After proving the generator property of $(A, D(A))$ we continue by deriving the associated semigroup and diffusion.

Theorem 2.6 *The linear operator $(A, D(A))$ generates a positive, strongly continuous, contraction semigroup $(T(t))_{t \geq 0}$ on $C_0([0, 1])$ of the form*

$$(T(t)f)(x) = \int_{[0,1]} f(y) \, p_t(x, dy)$$

for all $f \in C_0([0, 1])$, $t \geq 0$ and some Markov transition function $(p_t)_{t \geq 0}$ of a diffusion process $(X_t)_{t \geq 0}$.

Proof. The positivity of $-A^{-1}$ implies the positivity of the semigroup. By the Hahn-Banach Theorem there exists an extension $\tilde{T}(t)$ of $T(t)$ to $C([0, 1])$ for all $t \geq 0$. By the Riesz representation theorem we obtain

$$F(f) = (\tilde{T}(t)f)(x) = \int_{[0,1]} f(y) \, d\mu_{t,x}(y) = \int_{[0,1]} f(y) \, p_t(x, dy)$$

with a finite, positive Borel measure $p_t(x,\cdot)$ for all $f \in C([0,1])$ and $x \in [0,1]$.

$T(0) = Id$ implies

$$f(x) = (T(0)f)(x) = \int_{[0,1]} f(y) \, p_0(x, dy)$$

for all $f \in C_0([0,1])$ and $x \in [0,1]$. This means $p_0(x,\cdot) = \delta_x$.

The semigroup law $T(t+s) = T(t)T(s)$ implies the Chapman-Kolmogorov equation $p_{t+s}(x,B) = \int_{[0,1]} p_s(y,B) p_t(x,dy)$ for all measurable sets $B \subset [0,1]$ and $t,s \geq 0$.

The existence of a monotone sequence $(f_n)_{n\geq 1}$ in $C([0,1])$ with $\lim_{n\to\infty} f_n = \mathbb{1}_B$ pointwise for all measurable sets B and the monotone convergence theorem assure the measurability of $p_t(\cdot, B)$ for all measurable sets $B \subset [0,1]$.

Therefore there exists a Markov process $(X_t)_{t\geq 0}$ associated to the Markov transition function $(p_t)_{t\geq 0}$, i.e. the semigroup $(T(t))_{t\geq 0}$, its generator $(A, D(A))$ and finally Equation (2.18) with $p_t(x, B) = P(X_t \in B \mid X_0 = x)$ for all measurable sets $B \subset [0,1]$ and $t \geq 0$. □

Further this Markov transition function is stochastically continuous and has the Feller property. See e.g. Lamperti (1977) for details.

A useful identity that connects the semigroup and the inverse of its generator is given by the following proposition.

Proposition 2.3.10 *Let $(T(t))_{t\geq 0}$ be the semigroup generated by $(A, D(A))$ and $f \in C_0([0,1])$. Then*

$$\int_0^\infty (T(t)f)(x)\,dt = (-A^{-1}f)(x)$$

holds for all $x \in [0,1]$.

Proof. A proof can be found in Engel-Nagel (2000), Theorem II.1.10. □

We are now able to show that the Markov transition function $(p_t)_{t\geq 0}$ belonging to the semigroup $(T(t))_{t\geq 0}$ has a density with respect to the Lebesgue measure.

Proposition 2.3.11 *The Markov transition function* $(p_t)_{t\geq 0}$ *belonging to the semigroup* $(T(t))_{t\geq 0}$ *generated by* $(A, D(A))$ *has almost surely a density with respect to the Lebesgue measure.*

Proof. With Proposition 2.3.10 and Theorems 2.4 and 2.6 we obtain

$$\int_0^\infty \int_0^1 f(y) p_t(x, dy)\, dt = \int_0^\infty (T(t)f)(x)\, dt$$
$$= (-A^{-1}f)(x)$$
$$= -\frac{m(x)W(x)}{v_1} \int_0^x \frac{f(s)}{m(s)W(s)} ds - \frac{1}{v_1} \int_x^1 f(s)\, ds$$

for all $f \in C_0([0,1])$. This yields

$$f = 0 \; \lambda^1\text{-a.s.} \Rightarrow \int_0^1 f(y) p_t(x, dy) = 0 \text{ for } \lambda^1\text{-almost all } t.$$

It follows that $p_t(x, \cdot)$ has λ^1-almost surely a density. With λ^1 we denote the one-dimensional Lebesgue measure. $\square$

2.3.7 Proof of Theorem 2.3

After all this preparation a short proof of Theorem 2.3 can be given.

Proof. Consider the Diffusion process $(X_t)_{t\geq 0}$ associated to Equation (2.18). Let $(g_n^{[a,b[})_{n\geq 1} \subset D(A)$ be a sequence of positive functions with $g_n^{[a,b[} \nearrow \mathbb{1}_{[a,b[}$ for $n \to \infty$ pointwise for $0 \leq a < b \leq 1$. Then Fubini's theorem yields

$$E\left[\lambda^1(\{t : X_t \in [a,b[\})| X_0 = 0\right] = E\left[\int_0^\infty \mathbb{1}_{[a,b[}(X_t) dt \Big| X_0 = 0\right]$$
$$= \int_0^1 \int_0^\infty \mathbb{1}_{[a,b[}(x)\, dt\, p_t(0, dx)$$
$$= \int_0^\infty \int_0^1 \mathbb{1}_{[a,b[}(x) p_t(0, dx)\, dt.$$

With the monotone convergence theorem we further obtain

$$\begin{aligned}
\int_0^\infty \int_0^1 \mathbb{1}_{[a,b[}(x) p_t(0,dx)\, dt &= \int_0^\infty \int_0^1 (\lim_{n\to\infty} g_n^{[a,b[}(x)) p_t(0,dx)\, dt \\
&= \int_0^\infty \left\{ \lim_{n\to\infty} \int_0^1 g_n^{[a,b[}(x) p_t(0,dx) \right\} dt \\
&= \int_0^\infty \lim_{n\to\infty} (T(t) g_n^{[a,b[})(0)\, dt
\end{aligned}$$

while using the identities $\lim_{n\to\infty} g_n^{[a,b[} = \mathbb{1}_{[a,b[}$ for the sequence $(g_n^{[a,b[})_{n\geq 1}$ and $\int_0^1 f(x) p_t(0,dx) = (T(t)f)(x)$ for the semigroup $(T(t))_{t\geq 0}$ on $C_0([0,1])$. A repeated application of the monotone convergence theorem results in

$$\begin{aligned}
\int_0^\infty \lim_{n\to\infty} (T(t) g_n^{[a,b[})(0)\, dt &= \lim_{n\to\infty} \int_0^\infty (T(t) g_n^{[a,b[})(0)\, dt \\
&= \lim_{n\to\infty} (-A^{-1} g_n^{[a,b[})(0)
\end{aligned}$$

after replacing $\int_0^\infty (T(t)f)(x) dt$ with $(-A^{-1}f)(x)$ by Proposition 2.3.10. Theorem 2.4 and a final application of the monotone convergence theorem now yield the desired result

$$\begin{aligned}
\lim_{n\to\infty} (-A^{-1} g_n^{[a,b[})(0) &= \lim_{n\to\infty} \left[\tfrac{1}{v_1} \int_0^1 g_n^{[a,b[}(s)\, ds \right] \\
&= \tfrac{1}{v_1} \int_0^1 \lim_{n\to\infty} g_n^{[a,b[}(s)\, ds \\
&= \tfrac{1}{v_1} \int_0^1 \mathbb{1}_{[a,b[}(s)\, ds \\
&= \tfrac{b-a}{v_1}.
\end{aligned}$$

Here λ^1 denotes again the one-dimensional Lebesgue measure. $\square$

2.3.8 Examples

We consider some concrete examples.

Example 2.3.12 *Consider the operator*

$$Au(x) = (1-x)^\alpha u''(x) + (v_1 - \alpha(1-x)^{\alpha-1}) u'(x) \tag{2.45}$$

on $[0,1]$ for some $\alpha \geq 1$ and $v_1 \in (0,\infty)$. Then there exists a diffusion process associated to Equation (2.45) such that Danckwerts' law holds.

Proof. We have to verify that the conditions from Section 2.3.5 hold. We see that the diffusion function $D(x) = 2(1-x)^\alpha$ is continuous on $[0,1]$ and differentiable with derivative $D'(x) = -2\alpha(1-x)^{\alpha-1}$. It is positive on $[0,1)$ and zero at one. The derivatives at both boundaries exist with $D'(0) = -2\alpha$ and $D'(1) \leq 0$. Finally the drift function v is continuous and suffices $v(x) = v_1 - \alpha(1-x)^{\alpha-1} = \frac{1}{2}D'(x) + v_1$. □

The next example features a degenerate diffusion at the left boundary and therefore an entrance boundary.

Example 2.3.13 *Consider the operator*

$$Au(x) = x(1-x)u''(x) + (1 - 2x + v_1)u'(x) \tag{2.46}$$

on $[0,1]$ for some $v_1 \in (0, \infty)$. Then there exists a diffusion process associated to Equation (2.46) such that Danckwerts' law holds.

Proof. We have to verify that the conditions from Section 2.3.5 hold. We see that the diffusion function $D(x) = 2x(1-x)$ is continuous on $[0,1]$ and differentiable with derivative $D'(x) = 2(1 - 2x)$. It is positive on $(0,1)$ and zero at the boundaries. The derivatives at both boundaries exist with $D'(0) = 2$ and $D'(1) = -2$. Finally the drift function v is continuous and suffices $v(x) = 1 - 2x + v_1 = \frac{1}{2}D'(x) + v_1$. □

Not only polynomial functions but a wide variety of functions, e.g. exponential functions are possible candidates for the diffusion function.

Example 2.3.14 *Consider the operator*

$$Au(x) = (\exp(-\alpha x) - \exp(-\alpha))u''(x) + (v_1 - \alpha \exp(-\alpha x))u'(x) \tag{2.47}$$

on $[0,1]$ for some $\alpha \geq 1$ and $v_1 \in (0, \infty)$. Then there exists a diffusion process associated to Equation (2.47) such that Danckwerts' law holds.

Proof. We have to verify that the conditions from Section 2.3.5 hold. We see that the diffusion function $D(x) = 2(\exp(-\alpha x) - \exp(-\alpha))$ is continuous on $[0,1]$ and differentiable with derivative $D'(x) = -2\alpha \exp(-\alpha x)$. It is positive on $(0,1)$ and zero at one. The derivatives at both boundaries exist with $D'(0) = -2\alpha$ and $D'(1) = -2\alpha \exp(-\alpha) < 0$. Finally the drift function v is continuous and suffices $v(x) = v_1 - \alpha \exp(-\alpha x) = \frac{1}{2}D'(x) + v_1$. □

Chapter 3

Multiphase Processes

The choice of a concrete model is usually an important decision with a strong inherent effect. A former vague or inprecise notion of a process that shall be modeled is instantly transferred into the exact mathematical language which gives no room for different truths about the same object. It also carries the setting from which the examined process is now looked upon. Once a mathematical model is chosen all derived statements are true inside of the mathematical system, i.e. the chosen assumptions and considered objects. In addition certain techniques and results become available to deduce information from the setup. These techniques and results depend strongly on the chosen mathematical model and have to be taken into account when choosing the model. Normally a trade-off between generality of the setup and manageability of the model has to be made.

A stochastic process represents the most general model to describe the evolution of a randomly behaving system. It shows different faces for different state spaces, index sets and dependence structures. The differences in used techniques, existing results and available formulations between discrete and continuous processes are often huge. Further the dependence structure plays a starring role as well. Processes with independent random variables are quite different to handle than those which exhibit a dependence structure of some sort.

In this chapter we like to focus on socalled multiphase processes. Those multiphase processes are stochastic processes which usually appear in the context of creating a Markov process from a given stochastic process. Sometimes a model is not set up such that the stochastic process features a Markovian structure or an existing model loses its Markovian character while enlarging it to include additional properties. But it is still frequently possible to generate a Markov process which describes the process at hand and keeps the advantages that all tools concerning Markov processes are still available.

In the following we first present a motivating example from actual research. Afterwards we show how the conditions for Danckwerts' law to hold from Chapter 2 can be applied to a multiphase process. The last two sections link the discrete multiphase process to its continuous analogue. The existence of a diffusion limit is shown and a set of partial differential equations characterizing the diffusion limit is derived.

3.1 Bubbling Fluidized Bed Reactor Revisited

In an Example in Section 2.2.4 a Markov chain model for a bubbling fluidized bed has been introduced. This model consists of a Markov birth-death chain with an additional transition; a jump to state 1. This jump describes the extra transport upwards of particles caught in the wake of a rising gas bubble. Introducing such a jump to the top which amounts to admitting infinite velocity for those rising particles was certainly a model assumption of which feasibility and impact had still to be examined, although very good results had already been achieved with this approach, e.g. see Dehling, Hoffmann and Stuut (1999).

Recent experiments by Dechsiri et al. (2005) using a PET-camera show a quite fast rise of particles to the top but nonetheless one with finite velocity. This fact can clearly be observed in Figure 3.1. Therefore the model should be extended such that it exhibits this feature. Unfortunately a simple extension of the existing model is not possible since the necessary information if a particle is drifting and diffusing downwards with some velocity or moving upwards in the wake of a rising gas bubble with some other velocity is not preserved in the actual state, i.e. the particle's location in the reactor. Thus the process loses its Markovian character. Nevertheless an extension of the current model which preserves its Markovian character is possible by using a multiphase process.

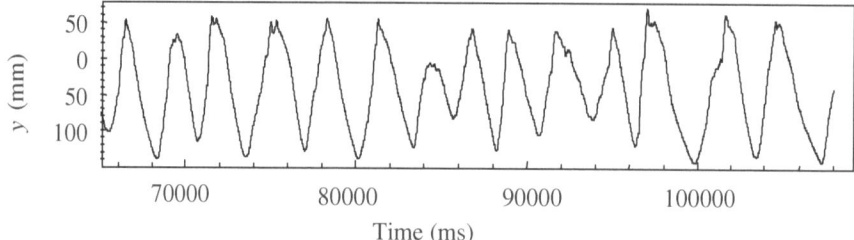

Figure 3.1: Time series of a tracer's particle height in bed

3.1.1 Original Model

Let us first recollect the basic facts about the original model, a Markov chain model for a bubbling fluidized bed reactor as introduced in Section 2.2.4 and shown in Figure 3.2.

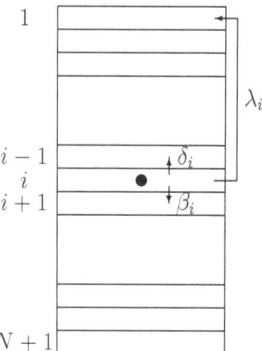

Figure 3.2: Discretized fluidized bed with arrows indicating all possible transition pathes for the particle located in cell i.

The bed is discretized in layers numbered from 1 to N and an exit (absorbing) state with

the label $N+1$. The transition matrix P consists of probabilities for $3 \leq i \leq N$:
$$\begin{aligned}
p_{i,i} &= \alpha_i = 1 - \beta_i - \delta_i - \lambda_i, \\
p_{i,i+1} &= \beta_i, \\
p_{i,i-1} &= \delta_i, \\
p_{i,1} &= \lambda_i
\end{aligned}$$
and
$$\begin{aligned}
p_{1,1} &= 1 - \beta_1, \\
p_{1,2} &= \beta_1, \\
p_{2,1} &= \delta_2 + \lambda_2, \\
p_{2,2} &= \alpha_2, \\
p_{2,3} &= \beta_2, \\
p_{N+1,N+1} &= 1,
\end{aligned}$$
at the boundaries. The object of examination is the particle's location at each timestep $n \geq 0$ denoted by the random variable $X_n : \Omega \to S$ with $S = \{1, 2, \ldots, N+1\}$. These random variables form the Markov chain $(X_n)_{n \geq 0}$.

3.1.2 Multiphase Model

To maintain the Markov property while incorporating a finite velocity for the upwards rising particles in the wake of gas bubbles we introduce a socalled second phase in the state space which leads to the following state space
$$S = \{1, 2, \ldots, N\} \times \{0, 1\} \cup \{(N+1, 0)\}.$$
We still keep track of the particle's location at each time step but we additionally know the particle's phase. As phases we distinguish here the ordinary downwards drift and diffusion and the fast rise upwards in the wake of a gas bubble. Thus the process is still Markovian whereas the projection on its first coordinate is not. With a slight abuse of notation we denote the multiphase system again by $(X_n)_{n \geq 0}$. The particle's location is encoded in the first and the two phases are encoded in the second entry of our state space variable. Now the transition probabilites are given as:
$$\begin{aligned}
p_{(i,k)(i-1,k)} &= \delta_i^{(k)}, \\
p_{(i,k)(i+1,k)} &= \beta_i^{(k)}, \\
p_{(i,k)(i,k)} &= \alpha_i^{(k)}, \\
p_{(i,k)(i,|k-1|)} &= \lambda_i^{(k)}
\end{aligned}$$

and suffice

$$\alpha_{N+1}^{(0)} = 1,$$
$$\delta_i^{(k)} + \beta_i^{(k)} + \alpha_i^{(k)} + \lambda_i^{(k)} = 1,$$

for $k \in \{0,1\}$, $1 \leq i \leq N$. Phase 0 models the downward flow as a usual drift and diffusion

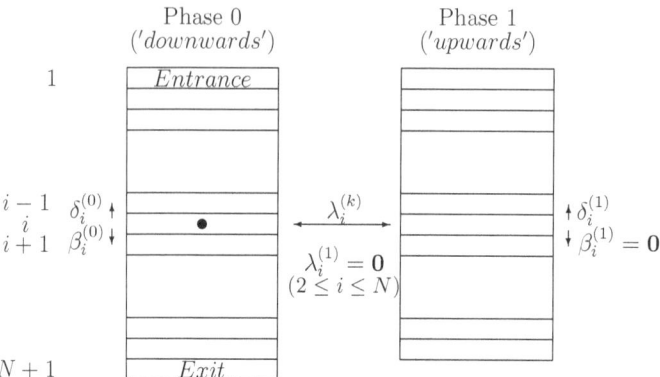

Figure 3.3: Multiphase system for a fluidized bed with arrows indicating all possible transition pathes for the particle located in cell i.

and phase 1 models the upward drift in the wake of a rising gas bubble by setting

$$\beta_i^{(1)} = 0$$

for $1 \leq i \leq N$. A value greater than zero for the parameters $\beta_i^{(1)}$ would model a different kind of drifting and diffusing flow in phase 1 than in phase 0 either upwards ($\beta_i^{(1)} < \delta_i^{(1)}$) or downwards ($\beta_i^{(1)} > \delta_i^{(1)}$) in each cell i. This behavior can be found in beds containing two different kinds of flow regimes, e.g. because of gulf streaming.

The change between the two phases, i.e. entering/leaving the wake of a gas bubble, is governed by $\lambda_i^{(k)}$. Setting

$$\lambda_i^{(1)} = 0$$

for $2 \leq i \leq N$ models that the upward phase will only be left at the top of the reactor. It is also possible to allow a change of phase in some specified or simply all cells to model that a particle rising in the wake of a gas bubble could leave this bubble in these specified locations. This would correspond to a fraction of material that is rising in the wake of gas bubbles to leave these bubbles and start the standard downwards drift and diffusion again. An example where this kind of modeling approach could gainfully be employed are e.g. fluidized bed reactors with baffles. The model is visualized in Figure 3.3.

It should be emphasized that this multiphase system is a Markov chain with all its advantages in contrast to the particle's location only which is the projection of the multiphase system on its first coordinate/entry and not Markovian. Enlarging the state space turns out to be a fruitful detours to gain the sought information about the behavior of the examined system. Even more information can be obtained using the multiphase system approach, e.g. it keeps track of the particle in each phase, i.e. the expected time of the particle in each phase can also be derived.

An even further expanded model with a state space of the form $\{1, 2, \ldots, N\} \times \{0, 1, 2\}$ from Dehling, Gottschalk and Hoffmann (2007) has been applied very successfully to the data presented above in Figure 3.1 as can be seen in Figure 3.4.

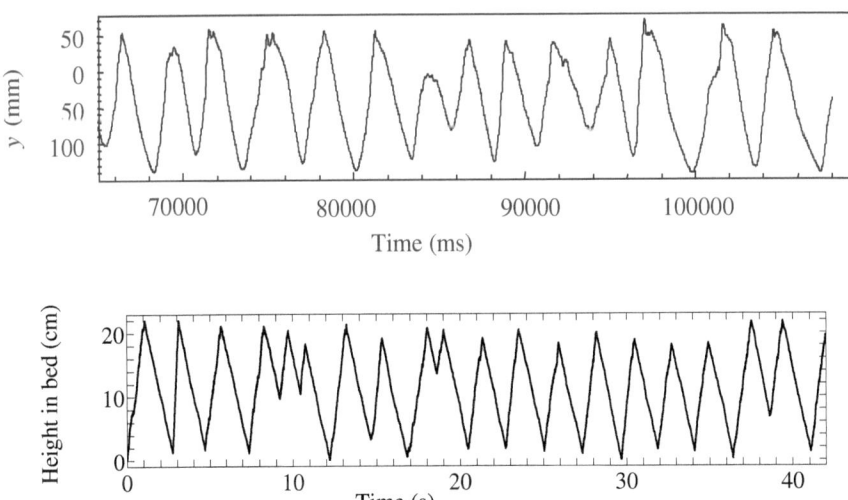

Figure 3.4: Time-series of the tracer particle's height in bed: experiment and simulation

For more information about this model and its validation with data we refer to Dehling, Gottschalk and Hoffmann (2007).

3.2 Danckwerts' Law Revisited

The validity of Danckwerts' law is unaffected of the type of used model be it a simple Markov chain model or a multiphase model. Hence we want to apply the results from Section 2.2 to a multiphase model as well. But before we give a precise definition of a multiphase process.

Definition 3.2.1 *Let $\tilde{S}$ be some space. A Markov process $(X_t)_{t\geq 0}$ with a state space S of cartesian product form, i.e.*

$$S = \tilde{S} \times \{1, 2, \ldots, K\}$$

*for some $K \in \mathbb{N} \setminus \{0, 1\}$ is called a **multiphase process**.*

Remark 3.2.2 *Technically each Markov chain can be interpreted as a multiphase process by renumbering of the states and identifying some of them. Nevertheless the representation of the state space can carry some information, e.g. in a multiphase process states with an equal coordinate might have common properties. Further the multiphase process has the advantage to bring along canonical subsets of the state space through projecting on some coordinate. Hence it is quite natural to distinguish between multiphase processes and Markov chains.*

Considering a Markov chain $(X_n)_{n\geq 0}$ and returning now to the basic results for Danckwerts' law to hold we find

1. $\sum_{j=1}^{N} p_{j,1} = 1 - v$,
2. $\sum_{j=1}^{N} p_{j,i} = 1$ for all $2 \leq i \leq N$,
3. $\sum_{j=1}^{N} p_{j,N+1} = v$,

as necessary conditions on the transfer probabilities. At the first glance these conditions only apply to a state space of the form $\{1, 2, \ldots, N+1\}$. In spite of that they can be used in the present case of a multiphase system with its product state space

$$S = \{1, 2, \ldots, N\} \times \{1, 2, \ldots, K\}.$$

We use renumbering of the state space to identify S with $\{1, 2, \ldots, NK\}$. For convenience we set $\tilde{N} + 1 = NK$ and assume the entrance in state 1 and the exit in state $\tilde{N} + 1$. Now the conditions from Section 2.2 can be applied. A renumbering of the form

$$\begin{cases} (1,1) & \longleftrightarrow & 1 \\ \vdots & \longleftrightarrow & \vdots \\ (1,K) & \longleftrightarrow & K \\ (2,1) & \longleftrightarrow & K+1 \\ \vdots & \longleftrightarrow & \vdots \\ (N,K) & \longleftrightarrow & NK \end{cases}$$

leads to

1. $\sum_{j=1}^{\tilde{N}+1} p_{j,1} = \sum_{m=n}^{N} \sum_{j=k}^{K} p_{(m,j),(n,k)} = 1 - v$,

2. $\sum_{j=1}^{\tilde{N}+1} p_{j,i} = \sum_{m=n}^{N} \sum_{l=k}^{K} p_{(m,l),(j_1,j_2)} = 1 \qquad \forall (j_1, j_2) \in S \setminus \{(1,1), (N,K)\}$ respectively $2 \leq i \leq \tilde{N}$,

3. $\sum_{j=1}^{\tilde{N}+1} p_{j,\tilde{N}+1} = \sum_{m=n}^{N} \sum_{l=k}^{K} p_{(m,l),(N,K)} = 1 + v$,

where the transition probabilities p_{ij} belong to the renumbered Markov chain associated to the transition probabilities $p_{(m,l)(j_1,j_2)}$. We note that we extended the summation in the conditions above to include the state $\tilde{N} + 1$ and used its absorbing nature, i.e. $p_{\tilde{N}+1,\tilde{N}+1} = 1$.

3.2.1 Danckwerts' Law for the Twophase Fluidized Bed Reactor Model

Using the approach described above the conditions for Danckwerts' law to hold for the two phase model for a bubbling fluidized bed reactor from Section 3.1.2 look as follows.

$$\begin{aligned} (i) \quad & \lambda_1^{(1)} + \delta_2^{(0)} + v = \beta_1^{(0)} + \lambda_1^{(0)}, \\ (ii) \quad & \delta_2^{(1)} + \lambda_1^{(0)} = \lambda_1^{(1)}, \\ (iii) \quad & \beta_{i-1}^{(0)} + \delta_{i+1}^{(0)} = \delta_i^{(0)} + \beta_i^{(0)} + \lambda_i^{(0)} \qquad \text{for all } 2 \leq i \leq N, \\ (iv) \quad & \delta_{i+1}^{(1)} + \lambda_i^{(0)} = \delta_i^{(1)} \qquad \text{for all } 2 \leq i \leq N \end{aligned}$$

The left hand sides in the conditions 1.-3., i.e. the inflows, are calculated as

$$\lambda_1^{(1)} + 1 - (\beta_1^{(0)} + \lambda_1^{(0)}) + \delta_2^{(0)} \qquad \text{for cell } (1,0),$$

$$\delta_2^{(1)} + \lambda_1^{(0)} + 1 - \lambda_1^{(1)} \qquad \text{for cell } (1,1),$$

$$\beta_{i-1}^{(0)} + \delta_{i+1}^{(0)} + 1 - (\delta_i^{(0)} + \beta_i^{(0)} + \lambda_i^{(0)}) \qquad \text{for all cells } (i,0) \text{ with } 2 \leq i \leq N,$$

$$\delta_{i+1}^{(1)} + 1 - \delta_i^{(1)} + \lambda_i^{(0)} \qquad \text{for all cells } (i,1) \text{ with } 2 \leq i \leq N.$$

Then equating via conditions 1.-3. above generates conditons (i)-(iv). These conditions (i)-(iv) show that for all cells in the interior the inflow equals the outflow, i.e. mass is preserved and the net-outflow from the exit equals v.

3.3 Diffusion Approximation

The principal result of the current chapter ensures the existence of a multiphase Markov process in continuous time and space as a limit of a sequence of multiphase Markov birth–death chains. In this context a characterization of the multiphase process is derived as well. We base our work on the diffusion approximation in Dabrowski and Dehling (1998) and extend it to multiphase processes.

3.3.1 The Processes

In this section we present the discrete processes and their approximation limit process which has continuous paths in the first coordinate. The discrete processes are constructed from a family of birth–death Markov chains and originate a sequence of Markov chains with increasing discretization fineness while the continuous process is composed of several diffusion processes with varying behavior.

We consider multiphase processes with state space and index set $[0, 1]$. To simplify the exposition we restrict our attention to multiphase processes with only two phases. The processes are governed by their drift, diffusion and phase transition functions.

drift function	v_k	:	$[0,1] \to \mathbb{R}$	$k \in \{1,2\}$
diffusion function	D_k	:	$[0,1] \to (0,\infty)$	$k \in \{1,2\}$
phase transition function	λ_{kl}	:	$[0,1] \to (0,\infty)$	$k \neq l; k,l \in \{1,2\}$

Later we need mild regularity properties of the drift, diffusion and phase transition functions and comment respectively assume these when they are used.

The Discrete Processes

The discrete processes consist of components which themselves are ordinary birth–death Markov chains with sensibly chosen transition probabilities as outlined in Section 2.3.1. These discrete processes are later brought into a continuous context via embedding and interpolating.

For each $N \in \mathbb{N} \setminus \{0\}$ and $k \in \{1,2\}$ the discrete birth–death Markov chains $(Y_n^N(k))_{n \in \mathbb{N}}$ are defined on the space $\{i\Delta \cdot 0 \leq i \leq N\}$ with $\Delta = \frac{1}{N}$. Their transition probabilities are given by parameters as in Section 2.3.1 as follow

$$p_{i,i-1}(k) = \delta_i^{(k)} = \frac{\varepsilon}{2\Delta^2} D_k(i\Delta) - \frac{\varepsilon}{2\Delta} v_k(i\Delta),$$

$$p_{i,i+1}(k) = \beta_i^{(k)} = \frac{\varepsilon}{2\Delta^2} D_k(i\Delta) + \frac{\varepsilon}{2\Delta} v_k(i\Delta),$$

$$p_{i,i}(k) = 1 - \delta_i^{(k)} - \beta_i^{(k)} = 1 - \frac{\varepsilon}{\Delta^2} D_k(i\Delta)$$

in the interior, i.e. $2 \leq i \leq N-1$, and absorbing and reflecting boundary conditions

$$p_{1,2}(k) = \beta_1^{(k)}$$

$$p_{1,1}(k) = 1 - \beta_1^{(k)}$$

$$p_{N,N}(k) = 1$$

at the endpoints of the state space. For the time rescaling parameter ε the identity $\varepsilon = \frac{\Delta^2}{\max_{x \in [0,1]} \max_{k \in \{1,2\}} D_k(x)}$ holds. These discrete Markov chains are then embedded in continuous time and space by setting

$$Y_t^N(k) = (i+1)Y_{i\varepsilon}^N(k) - iY_{(i+1)\varepsilon}^N(k) + \frac{t}{\varepsilon}(Y_{(i+1)\varepsilon}^N(k) - Y_{i\varepsilon}^N(k)) \qquad (3.1)$$

for $t \in [i\varepsilon, (i+1)\varepsilon] \cap [0,1]$ and $i \in \mathbb{N}$. Equation (3.1) constitutes a linear interpolation of the Markov chain after an appropriate rescaling of the time axis, and gives a stochastic process $(Y_t^N(k))_{t \in [0,1]}$ on $[0,1]$ with continuous paths for $k \in \{1,2\}$ and $N \in \mathbb{N} \setminus \{0\}$. These processes constitute the components in the construction of the multiphase processes $(Y_t^N)_{t \in [0,1]}$ later around Equation (3.5).

To each component we associate a stopping time using the phase transition functions λ_{kl}. These stopping times are also referred to as lifetimes. Hence we define the intensity functions of the discrete processes $(Y_t^N(k))_{t \in [0,1]}$ as

$$F_k^N(t) = \prod_{i=1}^{\lfloor \frac{t}{\varepsilon} \rfloor} \left(1 - \varepsilon \lambda_{kl}(Y_{(i-1)\varepsilon}^N(k))\right) \qquad (3.2)$$

for $k \neq l; k, l \in \{1,2\}$ and $t \in [0,1]$. The lifetimes $\tau^N(k)$ are then obtained as

$$\tau^N(k) = \min\{1, \inf\{t > 0 : F_k^N(t) \leq U\}\} \qquad (3.3)$$

for some $U[0,1]$–distributed random variable U independent of the process $(Y_t^N(k))_{t \in [0,1]}$ and $k \in \{1,2\}$.

We note that for $k \in \{1,2\}$ the process $(Y_t^N(k), k)_{t \in [0,1]}$ can be represented as

$$(Y_t^N(k), k)_{t \in [0,1]} = f((x,k); U) \qquad (3.4)$$

with some measurable function f. Here U denotes a $U[0,1]$–distributed random variable and (x,k) the starting point of $(Y_t^N(k), k)_{t \in [0,1]}$, i.e. $(Y_0^N(k), k) = (x,k)$.

Now we are able to define the twophase birth–death process by lifting the components defined above into a multiphase state space and glueing them together. Therefore let V_1^N denote a $U[0,1]$–distributed random variable and consider the processes with their associated lifetimes. The first pair of component and associated lifetime is given as

$$\left[(Y_{t,1}^N)_{t \in [0,1]}, \tau_1^N\right] = \left[f((0,1); V_1^N), \tau_{1,(0,1)}^N\right]$$

where the function f is defined in Equation (3.4) above. The associated lifetime τ_1^N is obtained as stated before in Equations (3.2) and (3.3) with a $U[0,1]$-distributed random variable U_1^N independent of V_1^N.

We continue inductively and use independent $U[0,1]$-distributed random variables V_i^N and U_i^N for $i \geq 1$. Given the n-th pair $\left[(Y_{t,n}^N)_{t\in[0,1]}, \tau_n^N\right]$ the $(n+1)$-th pair is defined via

$$\left[(Y_{t,n+1}^N)_{t\in[0,1]}, \tau_{n+1}^N\right] = \left[f((Y_{\tau_n^N,n}, 1 + \mathbb{1}_{\{n=0 \bmod 2\}}); V_{n+1}^N), \tau_{n+1,(Y_{\tau_n^N,n}, 1+\mathbb{1}_{\{n=0 \bmod 2\}})}^N\right] \quad (3.5)$$

again with the function f defined in Equation (3.4). We see that the construction above yields a sequence of processes whose first coordinates join together continuously, i.e.

$$Y_{0,n+1}^{N,1} = Y_{\tau_n^N,n}^{N,1},$$

where $Y_{t,i}^{N,1}$ denotes the first coordinate of $Y_{t,i}^N$, holds for all $n, N \geq 1$.

Here we like to point out that the random variable V_{n+1}^N driving the component process $(Y_{t,n+1}^N)_{t\in[0,1]}$ is independent of all previous random variables $V_1^N, \ldots, V_n^N$ and $U_1^N, \ldots, U_n^N$. The associated lifetime is obtained as stated above in Equations (3.2) and (3.3) with a random variable U_{n+1}^N independent of V_{n+1}^N and also independent of all previous random variables $V_1^N, \ldots, V_n^N$ and $U_1^N, \ldots, U_n^N$.

To reference the coordinates of the component processes we introduce the notion $Y_{t,n}^N = \left(Y_{t,n}^{N,1}, Y_{t,n}^{N,2}\right) \in [0,1] \times \{1,2\}$ for all $n \in \mathbb{N}$.

Finally the complete process is obtained by glueing the components selected by Equation (3.5) together. We obtain

$$Y_t^N = (0,1)\mathbb{1}_{\{t>\sum_{j=1}^\infty \tau_j^N\}} + \sum_{i=1}^\infty Y_{t-\sum_{j=1}^{i-1}\tau_j^N, i}^N \mathbb{1}_{\{\sum_{j=1}^{i-1}\tau_j^N \leq t < \sum_{j=1}^i \tau_j^N\}}$$

for all $t \in [0,1]$. The first term above $(0,1)\mathbb{1}_{\{t>\sum_{j=1}^\infty \tau_j^N\}}$ exists only to define the process on the whole space. Since the processes we discuss record no explosion, which is ensured by the boundedness of the phase transition functions λ_{kl} and is elaborated on later, it applies only to a set of measure zero.

In the following we refer to the processes $(Y_t^N)_{t\in[0,1]}$ and their components $(Y_{t,n}^N)_{t\in[0,1]}$ for $n \geq 1$ as the discrete process(es) in spite of the continuity of their paths. This makes sense since their natures are discrete while the continuity of their paths is artificially created by interpolating.

To give an intuitive description of the process $(Y_t^N)_{t\in[0,1]}$ we comment on its behavior. The process starts in the point $(0,1)$ and it behaves like a birth–death process with data v_1 and D_1 as introduced above until the first stopping time τ_1^N. Then it changes its phase, i.e second coordinate, into phase 2. Afterwards it behaves like a birth–death process with data v_2 and D_2. At the next stopping time its phase changes back into phase 1 and thereafter the procedure start over until finally the end of the time horizon $t = 1$ is reached. In between it constitutes a continuous process in the first coordinate when no explosion takes place. We stress that all pieces in the first coordinate in between phase transitions are approximating birth–death chains for the diffusion processes with the respective data v_k and D_k by e.g. Bhattacharya and Waymire (1990).

Further the phase transition functions $\lambda_{kl} : [0,1] \to [0,\infty)$ describe the phase transition rate such that the probability to change from phase k to phase l when at location $i\Delta$ in one timestep ε equals $\varepsilon \lambda_{kl}(i\Delta)$, i.e.

$$P\left(Y_{(j+1)\varepsilon}^N = (i\Delta, l) | Y_{j\varepsilon}^N = (i\Delta, k)\right) = \varepsilon \lambda_{kl}(i\Delta)$$

for all $1 \leq i \leq N$, $j \in \mathbb{N}$ such that $(j+1)\varepsilon \in [0,1]$ and $k \neq l; k, l \in \{1, 2\}$.

The Continuous Process

The continuous process $(Y_t)_{t\in[0,1]}$ associated to the discrete processes constructed above has values in its state space $[0,1] \times \{1,2\}$. Hence the word continuous here is used in the sense of continuous in the first coordinate. This is in accord with the notion of a multiphase process as introduced before.

We consider diffusion processes $(Y_t(k))_{t\in[0,1]}$ on $[0,1]$ with data v_k and D_k for $k \in \{1,2\}$. By Bhattacharya and Waymire (1990) those processes are approximation limits of their corresponding counterparts out of the previous section. To the diffusion processes we associate lifetimes similar as above. With the phase transition functions λ_{kl} the integrated intensity functions of the continuous processes $(Y_t(k))_{t\in[0,1]}$ are defined as

$$F_k(t) = \exp\left(-\int_0^t \lambda_{kl}(Y_s(k))\, ds\right) \tag{3.6}$$

for $k \neq l; k,l \in \{1,2\}$ and $t \in [0,1]$. The liftetimes $\tau(k)$ are then obtained as

$$\tau(k) = \min\{1, \inf\{t > 0 : F_k(t) \leq U\}\} \tag{3.7}$$

for some $U[0,1]$-distributed random variable U independent of the process $(Y_t(k))_{t\in[0,1]}$ and $k \in \{1,2\}$. Using the same random variables for the discrete and the continuous component processes, when generating their lifetimes, couples these lifetimes as shown in Lemma 3.3.5 below.

The twophase diffusion process can now be constructed by lifting the components above into a multiphase state space and glueing them together as in the discrete case. Having in mind that the continuous processes $(Y_t(k))_{t\in[0,1]}$ are approximation limits of the discrete processes $(Y_t^N(k))_{t\in[0,1]}$ we later establish the existence of versions of $(Y_t(k))_{t\in[0,1]}$ that are close to their approximating processes $(Y_t^N(k))_{t\in[0,1]}$ if the discretization fineness is large enough. Those continuous processes $(Y_t(k))_{t\in[0,1]}$ provide the origins for the components for the multiphase diffusion process.

Similiar to the discrete case we use a representation of the process $(Y_t(k), k)_{t\in[0,1]}$ for $k \in \{1,2\}$ as

$$(Y_t^N(k), k)_{t\in[0,1]} = g((x,k); U) \tag{3.8}$$

with some measurable function g, a $U[0,1]$-distributed random variable U and (x,k) the starting point of $(Y_t(k), k)_{t\in[0,1]}$, i.e. $(Y_0(k), k) = (x,k)$.

The first component and its associated lifetime are defined as

$$\left[(Y_{t,1})_{t\in[0,1]}, \tau_1\right] = \left[g((0,1); V_1), \tau_{1,(0,1)}\right]$$

where the function g is defined in Equation (3.8) and V_1 is a $U[0,1]$-distributed random variable. Its associated lifetime is obtained as stated before in Equations (3.6) and (3.7) with a $U[0,1]$-distributed random variable U_1 independent of V_1.

Analogue to our previous procedure we continue step by step. Given the n–th pair $\left[(Y_{t,n})_{t\in[0,1]}, \tau_n\right]$ the $(n+1)$–th pair is defined via

$$\left[(Y_{t,n+1})_{t\in[0,1]}, \tau_{n+1}\right] = \left[g((Y_{\tau_n,n}, 1 + 1\!\!1_{\{n=0\,mod\,2\}}); V_{n+1}), \tau_{n+1,(Y_{\tau_n,n}, 1+1\!\!1_{\{n=0\,mod\,2\}})}\right] \quad (3.9)$$

again with the function g defined in Equation (3.8). As in the discrete case the first coordinates of the component processes join together continuously and the random variable V_{n+1} is independent of all previous random variables $V_1, \ldots, V_n$ and $U_1, \ldots, U_n$. Its associated lifetime is again obtained as stated above in Equations (3.6) and (3.7) with a $U[0,1]$–distributed random variable U_{n+1} independent of V_{n+1} and all previous random variables $V_1, \ldots, V_n$ and $U_1, \ldots, U_n$.

Finally the complete process is obtained by glueing the components selected inductively by Equation (3.9) together. This yields

$$Y_t = (0,1) 1\!\!1_{\{t > \sum_{j=1}^\infty \tau_j\}} + \sum_{i=1}^\infty Y_{t - \sum_{j=1}^{i-1} \tau_j, i} 1\!\!1_{\{\sum_{j=1}^{i-1} \tau_j \leq t < \sum_{j=1}^i \tau_j\}}$$

for all $t \in [0,1]$. Again the first term above $(0,1) 1\!\!1_{\{t > \sum_{j=1}^\infty \tau_j\}}$ exists only to define the process on the whole space and by the lack of explosion it applies only to a set of measure zero. We maintain the analogue references to the coordinates of the continuous component processes with the notions $Y_{t,n} = \left(Y_{t,n}^1, Y_{t,n}^2\right) \in [0,1] \times \{1,2\}$ for all $n \in \mathbb{N}$.

An analoguous intuitive understanding of the continuous process $(Y_t)_{t\in[0,1]}$ as for the discrete process defined above $(Y_t^N)_{t\in[0,1]}$ holds. Here just the pieces in the first coordinate in between phase transitions which have been approximating birth–death chains for the diffusion processes in the discrete case have to be exchanged with their associated approximated diffusion processes. A similar fact for the interpretation of the phase transition functions λ_{kl} holds as well. We note that the stopping times of the continuous and the discrete process differ but should be close if the approximation fineness is small. That is discussed in Section 3.3.2 below.

3.3.2 Approximation

We show how the discrete processes $(Y_t^N)_{t\in[0,1]}$ defined above in Section 3.3.1 are approximations of the continuous process $(Y_t)_{t\in[0,1]}$ out of the same section. We begin by examining the lifetimes $(\tau_i)_{i\in\mathbb{N}}$ and $(\tau_i^N)_{i\in\mathbb{N}}$ of the component processes $((Y_{t,i})_{t\in[0,1]})_{i\in\mathbb{N}}$ and $((Y_{t,i}^N)_{t\in[0,1]})_{i\in\mathbb{N}}$ and showing that the number of components used in the construction of the processes above is stochastically finite. Furthermore we derive that the lifetimes are closely matched if their associated component processes are close. Afterwards we investigate the continuity properties for the paths of the continuous process and establish the closeness of the component processes if their starting points are close. This finally leads to the proof of the main theorem of this section where the approximating property of the processes $(Y_t^N)_{t\in[0,1]}$ to the process $(Y_t)_{t\in[0,1]}$ follows by combining all preparations made before.

Stopping Times

Here we give two results concerning the stopping times τ_i^N and τ_i above. At first we show that the number of phase transitions is stochastically finite and thereafter the closeness of the stopping times for the discrete and continuous processes is derived. We use ideas presented before in Dabrowski and Dehling (1998).

We set $\Lambda = \max_{k,l\in\{1,2\};k\neq l}||\lambda_{kl}||_\infty$ and assume throughout the entire section that $\varepsilon\Lambda < 1$ holds. Following are two lemmas which relate the stopping times to appropiate Γ-distributed random variables.

Lemma 3.3.1 *Let $(\tau_j)_{j\in\mathbb{N}}$ be defined as above in Equations (3.7) and (3.9), $K > 0$ and Z_K a $\Gamma(K,\Lambda)$-distributed random variable. Then stochastically $Z_K \leq \sum_{j=1}^K \tau_j$, i.e.*

$$P\left(\sum_{j=1}^K \tau_j \leq t\right) \leq P(Z_K \leq t)$$

holds for all $t \in [0,1]$.

Proof. We calculate

$$\begin{aligned}
P\left(\sum_{j=1}^{K} \tau_j \leq t\right) &= P\left(\tau_K \leq t - \sum_{j=1}^{K-1} \tau_j\right) \\
&= P\left(\exp\left(-\int_0^{t-\sum_{j=1}^{K-1}\tau_j} \lambda_{kl}(Y^1_{s,K}))\, ds\right) \leq U\right) \\
&\leq P\left(\exp\left(-(t - \sum_{j=1}^{K-1}\tau_j)\Lambda\right) \leq U\right) \\
&= P\left(\sum_{j=1}^{K-1} \tau_j - \tfrac{1}{\Lambda}\log U \leq t\right)
\end{aligned}$$

with some $U[0,1]$–distributed random variable U independent of $\{\tau_1,\ldots,\tau_{K-1}\}$. Iterating the scheme above we obtain

$$P\left(\sum_{j=1}^{K} \tau_j \leq t\right) \leq P\left(\sum_{j=1}^{K}(-\tfrac{1}{\Lambda}\log U_j) \leq t\right)$$

with independent $U[0,1]$–distributed random variables $U_1,\ldots,U_K$. By the transformation formula for densities we get

$$-\tfrac{1}{\Lambda}\log U_j \sim Exp(\Lambda)$$

for all $1 \leq j \leq K$ and therefore

$$\sum_{j=1}^{K}(-\tfrac{1}{\Lambda}\log U_j) \sim \Gamma(K, \Lambda)$$

to finish the proof. □

A version of Lemma 3.3.1 for the discrete processes holds as well.

Lemma 3.3.2 *Let $(\tau_j^N)_{j\in\mathbb{N}}$ be defined as above in Equations (3.3) and (3.5), $K > 0$ and Z_K^N a $\Gamma\left(K, -\frac{\log(1-\varepsilon\Lambda)}{\varepsilon}\right)$–distributed random variable. Then stochastically $Z_K^N \leq \sum_{j=1}^{K} \tau_j^N$, i.e.*

$$P\left(\sum_{j=1}^{K} \tau_j^N \leq t\right) \leq P(Z_K^N \leq t)$$

holds for all $t \in [0,1]$.

Proof. Analoguous to the proof of Lemma 3.3.1 we calculate

$$P\left(\sum_{j=1}^{K}\tau_j^N \leq t\right) = P\left(\tau_K^N \leq t - \sum_{j=1}^{K-1}\tau_j^N\right)$$

$$= P\left(\prod_{i=1}^{\left\lfloor\frac{t-\sum_{j=1}^{K-1}\tau_j^N}{\varepsilon}\right\rfloor}\left(1 - \varepsilon\lambda_{kl}(Y_{(i-1)\varepsilon,K}^{N,1})\right) \leq U\right)$$

$$\leq P\left((1-\varepsilon\Lambda)^{\frac{t-\sum_{j=1}^{K-1}\tau_j^N}{\varepsilon}} \leq U\right)$$

$$= P\left(\sum_{j=1}^{K-1}\tau_j^N + \frac{\varepsilon}{\log(1-\varepsilon\Lambda)}\log U \leq t\right)$$

with some $U[0,1]$-distributed random variable U independent of $\{\tau_1, \ldots, \tau_{K-1}\}$. Iterating the scheme above we obtain

$$P\left(\sum_{j=1}^{K}\tau_j^N \leq t\right) \leq P\left(\sum_{j=1}^{K}\frac{\varepsilon}{\log(1-\varepsilon\Lambda)}\log U_j \leq t\right)$$

with independent $U[0,1]$-distributed random variables $U_1, \ldots, U_K$. By the transformation formula for densities we get

$$\frac{\varepsilon}{\log(1-\varepsilon\Lambda)}\log U_j \sim Exp\left(-\frac{\log(1-\varepsilon\Lambda)}{\varepsilon}\right)$$

for all $1 \leq j \leq K$ and therefore

$$\sum_{j=1}^{K}\frac{\varepsilon}{\log(1-\varepsilon\Lambda)}\log U_j \sim \Gamma\left(K, -\frac{\log(1-\varepsilon\Lambda)}{\varepsilon}\right)$$

to finish the proof. □

With Lemmas 3.3.1 and 3.3.2 we can prove the finiteness of the number of phase transitions on a stochastically arbitrarily large set. This is made precise in the following proposition.

Proposition 3.3.3 *For all $\delta > 0$ there exists a $K \in \mathbb{N}\setminus\{0\}$ such that the number of phase transitions $\sum_{j=1}^{\infty}\mathbb{1}_{\{\sum_{i=1}^{j}\tau_i^N \leq 1\}}$ respectively $\sum_{j=1}^{\infty}\mathbb{1}_{\{\sum_{i=1}^{j}\tau_i \leq 1\}}$ for the processes $(Y_t^N)_{t\in[0,1]}$ and $(Y_t)_{t\in[0,1]}$ is bounded by K on a set of measure at least $1 - \delta$, i.e.*

$$P\left(\max\left\{\sum_{j=1}^{\infty}\mathbb{1}_{\{\sum_{i=1}^{j}\tau_i^N \leq 1\}}, \sum_{j=1}^{\infty}\mathbb{1}_{\{\sum_{i=1}^{j}\tau_i \leq 1\}}\right\} \leq K\right) \geq 1 - \delta$$

holds for all $N \in \mathbb{N}\setminus\{0\}$.

Proof. We consider the probability

$$P\left(\max\left\{\sum_{j=1}^{\infty} \mathbb{1}_{\{\sum_{i=1}^{j} \tau_i^N \leq 1\}}, \sum_{j=1}^{\infty} \mathbb{1}_{\{\sum_{i=1}^{j} \tau_i \leq 1\}}\right\} > K\right)$$
$$\leq P\left(\sum_{j=1}^{\infty} \mathbb{1}_{\{\sum_{i=1}^{j} \tau_i^N \leq 1\}} > K\right) + P\left(\sum_{j=1}^{\infty} \mathbb{1}_{\{\sum_{i=1}^{j} \tau_i \leq 1\}} > K\right)$$
$$= P\left(\sum_{i=1}^{K+1} \tau_i^N \leq 1\right) + P\left(\sum_{i=1}^{K+1} \tau_i \leq 1\right)$$

and show that both terms in the last line become arbitrarily small when K tends to infinity. The equality in the the third line holds since $\sum_{i=1}^{l} \tau_i$ and $\sum_{i=1}^{l} \tau_i^N$ are monotone increasing functions in l.

By applying Lemmas 3.3.1 and 3.3.2 we obtain

$$\begin{aligned} P\left(\sum_{i=1}^{K+1} \tau_i^N \leq 1\right) &\leq P(Z_{K+1}^N \leq 1) \\ P\left(\sum_{i=1}^{K+1} \tau_i \leq 1\right) &\leq P(Z_{K+1} \leq 1) \end{aligned} \qquad (3.10)$$

with $Z_{K+1}^N \sim \Gamma\left(K+1, -\frac{\log(1-\varepsilon\Lambda)}{\varepsilon}\right)$ and $Z_{K+1} \sim \Gamma(K+1, \Lambda)$. The probability distribution function of a $\Gamma(k, \lambda)$–distributed random variable Z has the form

$$F_Z(t) = P(Z \leq t) = 1 - e^{-\lambda t} \sum_{i=0}^{k-1} \frac{(t\lambda)^i}{i!}$$

for $t \geq 0$ and $k \in \mathbb{N} \setminus \{0\}$. Therewith we see that both right hand sides in the Inequalities (3.10) above tend to 0 for $K \to \infty$.

We finally note that K can be chosen independent of N since $-\frac{\log(1-\varepsilon\Lambda)}{\varepsilon} \to \Lambda$ for $\varepsilon \to 0$, i.e. $N \to \infty$. $\square$

The next step is to prove that closeness of the processes $(Y_{t,i})_{t \in [0,1]}$ and $(Y_{t,i}^N)_{t \in [0,1]}$ implies closeness of the stopping times τ_i and τ_i^N. Therefore we take a look at the intensity functions first. The following lemmas, proposition and proofs are closely related to those presented in Dabrowski and Dehling (1998). Nevertheless we adapt and expand those results to the case at hand. Hence we give complete proofs.

Again we present two lemmas first which constitute the fundamental parts of the proof of the second main result of this section which is given thereafter.

Lemma 3.3.4 Let $f, g : [0,1] \to [0,1]$ be measurable functions and λ_{kl} continuously differentiable for $k, l \in \{1,2\}; k \neq l$. Set $G_k^N(t) = \prod_{i=1}^{\lfloor \frac{t}{\varepsilon} \rfloor} (1 - \varepsilon \lambda_{kl}(g((i-1)\varepsilon)))$, $G_k(t) = \exp\left(-\int_0^t \lambda_{kl}(f(s))\, ds\right)$ and $\Lambda' = \max_{k,l \in \{1,2\}; k \neq l} ||\lambda'_{kl}||_\infty$. Then the following inequality holds

$$||G_k - G_k^N||_\infty \leq \varepsilon(\Lambda + \Lambda^2) + \Lambda' ||f - g||_\infty$$

for $k \in \{1, 2\}$. We recollect the definition of Λ as $\Lambda = \max_{k,l \in \{1,2\}; k \neq l} ||\lambda_{kl}||_\infty$.

Proof. At first we slice the interval $[0, 1]$ in commensurate intervals of lenght ε and then we lead the problem back to these. We estimate

$$\left| \exp\left(-\int_{(i-1)\varepsilon}^{i\varepsilon} \lambda_{kl}(f(s))\, ds\right) - (1 - \varepsilon \lambda_{kl}(g((i-1)\varepsilon))) \right|$$

$$= \left| \exp\left(-\int_{(i-1)\varepsilon}^{i\varepsilon} \lambda_{kl}(f(s))\, ds\right) - \left(1 - \int_{(i-1)\varepsilon}^{i\varepsilon} \lambda_{kl}(g((i-1)\varepsilon))\, ds\right) \right|$$

$$= \left| \exp\left(-\int_{(i-1)\varepsilon}^{i\varepsilon} \lambda_{kl}(f(s))\, ds\right) - \left(1 - \int_{(i-1)\varepsilon}^{i\varepsilon} \lambda_{kl}(g((i-1)\varepsilon))\, ds\right) \right.$$
$$\left. + \int_{(i-1)\varepsilon}^{i\varepsilon} \lambda_{kl}(f(s))\, ds - \int_{(i-1)\varepsilon}^{i\varepsilon} \lambda_{kl}(f(s))\, ds \right|$$

$$\leq \left(\int_{(i-1)\varepsilon}^{i\varepsilon} \lambda_{kl}(f(s))\, ds\right)^2 + \int_{(i-1)\varepsilon}^{i\varepsilon} |\lambda_{kl}(g((i-1)\varepsilon)) - \lambda_{kl}(f(s))|\, ds$$

$$\leq \varepsilon^2 \Lambda^2 + \varepsilon \Lambda' ||g - f||_\infty$$

for all $1 \leq i \leq \lfloor \frac{1}{\varepsilon} \rfloor$. The first estimate above uses the Taylor series expansion of the exponential function while the last estimate holds by the mean value theorem. Now we apply the inequality $|\prod_{i=1}^n a_i - \prod_{i=1}^n b_i| \leq \sum_{i=1}^n |a_i - b_i|$ which is valid for $a_i, b_i \in [-1, 1]$. We obtain while using the estimate above

$$|G_k(t) - G_k^N(t)| \leq \left| \exp\left(-\int_{\lfloor \frac{t}{\varepsilon} \rfloor \varepsilon}^t \lambda_{kl}(f(s))\, ds\right) \prod_{i=1}^{\lfloor \frac{t}{\varepsilon} \rfloor} \exp\left(-\int_{(i-1)\varepsilon}^{i\varepsilon} \lambda_{kl}(f(s))\, ds\right) \right.$$
$$\left. - \prod_{i=1}^{\lfloor \frac{t}{\varepsilon} \rfloor} (1 - \varepsilon \lambda_{kl}(g((i-1)\varepsilon))) \right|$$

$$\leq \sum_{i=1}^{\lfloor \frac{t}{\varepsilon} \rfloor} \left| \exp\left(-\int_{(i-1)\varepsilon}^{i\varepsilon} \lambda_{kl}(f(s))\, ds\right) - (1 - \varepsilon \lambda_{kl}(g((i-1)\varepsilon))) \right|$$
$$+ \left| \exp\left(-\int_{\lfloor \frac{t}{\varepsilon} \rfloor \varepsilon}^t \lambda_{kl}(f(s))\, ds\right) - 1 \right|$$

$$\leq \lfloor \tfrac{1}{\varepsilon} \rfloor (\varepsilon^2 \Lambda^2 + \varepsilon \Lambda' ||g - f||_\infty) + \varepsilon \Lambda$$

for all $t \in [0, 1]$ since $0 \leq s - \lfloor \frac{s}{\varepsilon} \rfloor \varepsilon \leq \varepsilon$ holds for all $s > 0$. This concludes the proof. $\square$

After deducing a result for the intensity functions in Lemma 3.3.4 we consider the stopping times in the second lemma and relate them to their intensity functions.

Lemma 3.3.5 *Consider a $U[0,1]$-distributed random variable U, intensity functions F_k and F_k^N as defined above in Equations (3.2) and (3.6) for $N \geq 1$, $k \in \{1,2\}$ and some $\delta > 0$. Set $\Lambda^- = \min_{k,l \in \{1,2\}; k \neq l} \min_{x \in [0,1]} \lambda_{kl}(x)$ for continuously differentiable phase transition functions λ_{kl}. Then $||F_k - F_k^N||_\infty \leq \delta$ implies*

$$|\tau(k) - \tau^N(k)| \leq \frac{e^\Lambda}{\Lambda^-}\left(\delta + \frac{1}{N}\right)$$

for the stopping times $\tau(k)$ and $\tau^N(k)$ associated to F_k and F_k^N which use the same random variable U.

Proof. For the derivative of the intensity function F_k we obtain

$$|F_k'(t)| \geq \Lambda^- F_k(t) \geq \Lambda^- e^{-\Lambda} \geq 0 \qquad (3.11)$$

for all $t \in [0,1]$. Now using the mean value theorem for F_k yields

$$\begin{aligned}\tau(k) - \tau^N(k) &= \tfrac{1}{F_k'(\xi)}\left(F_k(\tau(k)) - F_k(\tau^N(k))\right) \\ &= \tfrac{1}{F_k'(\xi)}\left(F_k(\tau(k)) - F_k^N(\tau^N(k)) + F_k^N(\tau^N(k)) - F_k(\tau^N(k))\right)\end{aligned} \qquad (3.12)$$

for some intermediate point ξ between $\tau(k)$ and $\tau^N(k)$. By combining Estimate (3.11) and Equation (3.12) we get

$$\begin{aligned}|\tau(k) - \tau^N(k)| &\leq \tfrac{e^\Lambda}{\Lambda^-}\left(|F_k(\tau(k)) - F_k^N(\tau^N(k))| + |F_k^N(\tau^N(k)) - F_k(\tau^N(k))|\right) \\ &\leq \tfrac{e^\Lambda}{\Lambda^-}\left(\tfrac{1}{N} + \delta\right)\end{aligned}$$

since $|F_k(\tau(k)) - F_k^N(\tau^N(k))| \leq \tfrac{1}{N}$ holds by the definition of the stopping times in Equations (3.3) and (3.7) since the same random variable U is used to determine both stopping times. □

With both lemmas above we derive the closeness of the stopping times for sufficient approximation quality.

Proposition 3.3.6 *Let $\delta > 0$ and consider processes $(Y_t^N(k))_{t\in[0,1]}$ and $(Y_t(k))_{t\in[0,1]}$ for $k \in \{1,2\}$ introduced in Section 3.3.1 with associated intensity functions F_k^N and F_k, stopping times τ_k^N and τ_k using a common $U[0,1]$-distributed random variable U. Then there exist $N_0 \in \mathbb{N}$ and $\delta_1 > 0$ such that*

$$\sup_{t\in[0,1]} |Y_t^N(k) - Y_t(k)| \leq \delta_1$$

for $N \geq N_0$ implies

$$|\tau(k) - \tau^N(k)| < \delta$$

for $N \geq N_0$.

Proof. This a direct consequence of Lemmas 3.3.4 and 3.3.5. □

Uniform Continuity and Approximation Properties of the Component Processes

The uniform continuity of the paths $t \to Y_t^1$ with $t \in [0,1]$ is later used to control the approximation on parts of the index set $[0,1]$ where the two processes $(Y_t)_{t\in[0,1]}$ and $(Y_t^N)_{t\in[0,1]}$ do not share the same second coordinate, i.e. phase. This is prepared for in the following proposition.

Lemma 3.3.7 *For all $\delta > 0$ there exists $n_0 \in \mathbb{N} \setminus \{0\}$ such that*

$$P\left(\sup_{|t-s|<\frac{1}{n}} |Y_t^1 - Y_s^1| > \delta \right) < \delta$$

holds for all $n \geq n_0$.

Proof. By the definition of $(Y_t)_{t\in[0,1]}$ and Proposition 3.3.3 the paths of the process $(Y_t^1)_{t\in[0,1]}$ are continuous on a set with measure one. On the compact set $[0,1]$ we therefore obtain uniform convergence and thus almost sure convergence of $a_n = \sup_{|t-s|<\frac{1}{n}} |Y_t^1 - Y_s^1|$

to zero. This further implies convergence in probability to zero and is equivalent to the claim of the lemma. □

In the following we need a version of Lemma 3.3.7 for the component processes $(Y^1_{t,i})_{t\in[0,1]}$, too.

Corollary 3.3.8 *Let $K \geq 1$, $\delta > 0$ and consider for $1 \leq i \leq K$ the component processes $(Y^1_{t,i})_{t\in[0,1]}$ of the process $(Y^1_t)_{t\in[0,1]}$. Then there exists $n_0 \in \mathbb{N} \setminus \{0\}$ such that*

$$P\left(\max_{1\leq j\leq K}\left\{\sup_{|t-s|<\frac{1}{n}}|Y^1_{t,j}-Y^1_{s,j}|\right\}>\delta\right)<\delta$$

holds for all $n \geq n_0$.

Proof. The same arguments as in the proof of Lemma 3.3.7 apply. □

The main tool to prove the approximation is given by results from Bhattacharya and Waymire (1990) and Rogers and Williams (1987). These yield that the birth–death Markov chain components of the discrete process $(Y^N_t)_{t\in[0,1]}$ are approximations of the associated components of the continuous process $(Y_t)_{t\in[0,1]}$. The Strassen–Dudley Theorem assures the existence of versions that are close in probability. We elaborate on this in Proposition 3.3.9. But before we state the Dudley–Strassen Theorem for a quick reference.

Theorem (Strassen–Dudley) *Let (S,d) be a complete separable metric space and μ and ν probability measures. Then the following two conditions are equivalent*

(i) *$\rho(\mu,\nu) \leq \varepsilon$ for the Prohorov metric ρ.*
(ii) *On some probability space there exist random variables X and Y with values in S and laws $\mathcal{L}(X) = \mu$ and $\mathcal{L}(Y) = \nu$, such that $P(d(X,Y) \geq \varepsilon) \leq \varepsilon$ holds.*

Proof. We see that the implication (ii) $\Rightarrow$ (i) is quite trivial in contrast to the other implication since for any measurable set A if $X \in A$ and $d(X,Y) < \varepsilon$ we have that $Y \in A^\varepsilon = \{y \in S : \text{there exists } x \in A \text{ with } d(x,y) < \varepsilon\}$ and thus

$$\mathcal{L}(X)(A) = P(X \in A) \leq P(Y \in A^\varepsilon) + \varepsilon = \mathcal{L}(Y)(A^\varepsilon) + \varepsilon$$

holds. Hence we obtain

$$\rho(\mathcal{L}(X), \mathcal{L}(Y)) = \rho(\mu, \nu) \leq \varepsilon$$

for the Prohorov metric ρ. In opposite to the implication $(ii) \Rightarrow (i)$ the other implication is a deep result and for a proof we refer to Dudley (2002), Theorem 11.6.2. □

Applying the Strassen–Dudley Theorem usually yields a version of a random variable X, i.e. another random variable Y with the same state space and the same law, i.e. $\mathcal{L}(X) = \mathcal{L}(Y)$. To limit the number of used notations we abstain from introducing new notations for these versions.

Sometimes it is advantageous to keep a certain random variable instead of using a version. In this case the following result by A.V. Skorohod comes in handy.

Theorem (Skorohod) *Let S and T be complete separable metric spaces, μ a probability measure on $S \times T$ with marginal μ_1 on S and $X : \Omega \to S$ a random variable with law $\mathcal{L}(X) = \mu_1$. Assume that there exists a $U[0,1]$-distributed random variable $U : \Omega \to [0,1]$ which is independent of X. Then there exists a measurable function $f : S \times [0,1] \to T$ such that $\mathcal{L}(X, f(X,U)) = \mu$ holds.*

Proof. We refer to the proof of Skorohod (1976) Theorem 1. □

Now it is time to present the mentioned proposition.

Proposition 3.3.9 *Consider $\delta > 0$, $k \in \{1,2\}$, discrete processes $(Y_t^N(k))_{t \in [0,1]}$ and continuous ones $(Y_t(k))_{t \in [0,1]}$ as introduced above with a continuous diffusion function D_k and a continuous drift function v_k. Let $Y_0^N(k) \to Y_0(k)$ hold in probability. Then there exist versions of the processes $(Y_t^N(k))_{t \in [0,1]}$ and some $N_0 \geq 1$ such that*

$$P\left(\sup_{t \in [0,1]} |Y_t^N(k) - Y_t(k)| > \delta\right) < \delta$$

holds for all $N \geq N_0$.

Proof. To begin with we fix starting points $y_1, y_2 \in [0,1]$ with $P(y_1 = Y_0^N(k)) = 1$ and $P(y_2 = Y_0(k)) = 1$ and use the notations $(Y_t^N(k, y_1))_{t \in [0,1]} = (Y_t^N(k))_{t \in [0,1]}$ and $(Y_t(k, y_2))_{t \in [0,1]} = (Y_t(k))_{t \in [0,1]}$. We estimate

$$\rho((Y_t^N(k, y_1))_{t \in [0,1]}, (Y_t(k, y_2))_{t \in [0,1]}) \leq \rho((Y_t^N(k, y_1))_{t \in [0,1]}, (Y_t(k, y_1))_{t \in [0,1]}) \\ + \rho((Y_t(k, y_1))_{t \in [0,1]}, (Y_t(k, y_2))_{t \in [0,1]}) \quad (3.13)$$

in the Prohorov metric ρ. By Bhattacharya and Waymire (1990) Theorem V.4.1 the discrete processes $(Y_t^N(k, y_1))_{t \in [0,1]}$ converge in distribution to the continuous processes $(Y_t(k, y_1))_{t \in [0,1]}$ for identical starting points. Theorem V.23.8 in Rogers and Williams (1987) yields that the distribution of the limit process $(Y_t(k, y_2))_{t \in [0,1]}$ depends continuously on the starting point y_2 of the process. Hence we obtain by Equation (3.13)

$$\rho((Y_t^N(k, y_1))_{t \in [0,1]}, (Y_t(k, y_2))_{t \in [0,1]}) \to 0 \quad (3.14)$$

for $N \to \infty$ and $y_1 \to y_2$.

Consider now random starting points and let $A = \{\sup_{t \in [0,1]} |Y_t^N(k) - Y_t(k)| > \delta\}$ and $B = \{|Y_0^N(k) - Y_0(k)| \leq \delta_B\}$. We calculate

$$P\left(\sup_{t \in [0,1]} |Y_t^N(k) - Y_t(k)| > \delta\right) = P(A \cap B) + P(A \cap B^c) \\ \leq \int E[\mathbb{1}_{A \cap B} \mid Y_0^N(k) = y_1, Y_0(k) = y_2] dQ(y_1, y_2) + P(B^c) \\ = \int_B P(A \mid Y_0^N(k) = y_1, Y_0(k) = y_2) dQ(y_1, y_2) + P(B^c)$$

where Q denotes the joint measure of $Y_0^N(k)$ and $Y_0(k)$. By Equation (3.14) there exist $\delta' > 0$ and $N_0 \geq 1$ such that

$$\rho((Y_t^N(k, y_1))_{t \in [0,1]}, (Y_t(k, y_2))_{t \in [0,1]}) < \frac{\delta}{2}$$

holds for all $N \geq N_0$ and $|y_1 - y_2| < \delta'$. Then the Strassen–Dudley Theorem and the Theorem of Skorohod yield versions of $(Y_t^N(k))_{t \in [0,1]}$ with

$$P(A \mid |Y_0^N(k) - Y_0(k)| < \delta') < \frac{\delta}{2}$$

for all $N \geq N_0$. Now choose $\delta_B = \frac{\delta'}{2}$ and $N_1 \geq N_0$ such that $P(B^c) < \frac{\delta}{2}$. This finally yields

$$P\left(\sup_{t \in [0,1]} |Y_t^N(k) - Y_t(k)| > \delta\right) \leq \int_B P(A \mid Y_0^N(k) = y_1, Y_0(k) = y_2) dQ(y_1, y_2) + P(B^c) \\ < \frac{\delta}{2} + \frac{\delta}{2}$$

since $|y_1 - y_2| < \delta'$ holds on B. □

One Step Approximation

In the proof of Proposition 3.3.13 Lemma 3.3.11 is used which is based on the preliminary Lemma 3.3.10. Lemma 3.3.10 allows to control the approximation when matching the components of the discrete processes to their associated components of the continuous process. Special attention is given to control the gaps between the lifetimes. Continuity properties of the continuous process are derived which are useful to bound the approximation error when both processes have different phases, i.e. second coordinates. This one step approximation below is the basis of the proof of Lemma 3.3.11.

Lemma 3.3.10 *Let $\delta > 0$ and $i \geq 1$. Then there exist $\delta' > 0$, $N_0 \geq 1$ and versions of $(Y_{t,i}^{N,1})_{t \in [0,1]}$ such that*

$$P\left(C_{1,i}(\delta) \cap C_{2,i}(\delta) \cap C_{3,i}(\delta) \mid B_{\delta',i}(N)\right) \geq 1 - \delta \quad (3.15)$$

holds with

$$C_{1,i}(\delta) = \left\{\sup_{t \in [0,1]} |Y_{t,i}^{N,1} - Y_{t,i}^1| \leq \delta\right\}$$
$$C_{2,i}(\delta) = \left\{\sup_{|t-s| < |\tau_i - \tau_i^N| + \delta'} |Y_t^1 - Y_s^1| + \sup_{|t-s| < |\tau_i - \tau_i^N| + \delta'} |Y_{t,i}^1 - Y_{s,i}^1| \leq \delta\right\}$$
$$C_{3,i}(\delta) = \left\{|\tau_i - \tau_i^N| \leq \delta\right\}$$
$$B_{\delta',i}(N) = \left\{|Y_{0,i}^{N,1} - Y_{0,i}^1| \leq \delta'\right\}$$

for all $N \geq N_0$. We note that the Lemma stays true if $\delta' > 0$ becomes smaller.

Proof. We show that

$$P\left(C_{1,i}^c(\delta) \cup C_{2,i}^c(\delta) \cup C_{3,i}^c(\delta) \mid B_{\delta',i}(N)\right) < \delta.$$

Let $\delta > 0$ and denote by $0 < \delta_{cont}$ a number such that

$$P\left(\sup_{|t-s| < \delta_{cont}} |Y_t^1 - Y_s^1| + \sup_{|t-s| < \delta_{cont}} |Y_{t,i}^1 - Y_{s,i}^1| > \frac{\delta}{2} \mid B_{\delta',i}(N)\right) < \frac{\delta}{2} \quad (3.16)$$

holds. The existence of such a $\delta_{cont} > 0$ is ensured by Lemma 3.3.7 and Corollary 3.3.8. We note that the proofs of Lemma 3.3.7 and Corollary 3.3.8 also hold for the measure $P(\cdot \mid B_{\delta',i}(N))$ for all $\delta' > 0$ and $N \geq 1$ since the considered processes are all continuous on a set of measure one.

Subsequently our focus drifts towards the lifetimes. We have seen that the lifetimes are close if the processes are close and N is large. Denote by δ_{lt} and N_{lt} the numbers that assure

$$|\tau_i - \tau_i^N| < \min\left\{\frac{\delta_{cont}}{2}, \frac{\delta}{2}\right\} \tag{3.17}$$

if $\sup_{t\in[0,1]} |Y_{t,i}^{N,1} - Y_{t,i}^1| \leq \delta_{lt}$ and $N \geq N_{lt}$ hold. The numbers δ_{lt} and N_{lt} exist by Proposition 3.3.6.

By Proposition 3.3.9 there exist for $\delta_{ver} = \min\{\delta_{lt}, \delta\}$ a $\frac{\delta_{cont}}{2} > \delta' > 0$ and a $N_0 \geq N_{lt}$ such that there exist versions of $(Y_{t,i}^{N,1})_{t\in[0,1]}$ that suffice

$$P\left(\sup_{t\in[0,1]} |Y_{t,i}^{N,1} - Y_{t,i}^1| \leq \delta_{ver} \mid B_{\delta'}(N)\right) \geq 1 - \delta_{ver} \tag{3.18}$$

for all $N \geq N_0$.

Fix $N \geq N_0$ and obtain on $B_{\delta',i}(N)$

$$C_{1,i}(\delta_{ver}) \subset C_{3,i}\left(\min\left\{\frac{\delta_{cont}}{2}, \frac{\delta}{2}\right\}\right) \tag{3.19}$$

by Equations (3.18) and (3.17). Further Equations (3.17) implies

$$C_{3,i}\left(\min\left\{\frac{\delta_{cont}}{2}, \frac{\delta}{2}\right\}\right) \cap \tilde{C}_{2,i}\left(\delta_{cont}, \frac{\delta}{2}\right) \subset C_{2,i}\left(\frac{\delta}{2}\right) \tag{3.20}$$

with

$$\tilde{C}_{2,i}(\delta_1, \delta_2) = \left\{\sup_{|t-s|<\delta_1} |Y_t^1 - Y_s^1| + \sup_{|t-s|<\delta_1} |Y_{t,i}^1 - Y_{s,i}^1| \leq \delta_2\right\}.$$

Since $\delta_1 \leq \delta_2$ implies $C_{m,i}(\delta_1) \subset C_{m,i}(\delta_2)$ for $m \in \{1,2,3\}$ we conclude

$$P\left(C_{1,i}^c(\delta) \cup C_{2,i}^c(\delta) \cup C_{3,i}^c(\delta) \mid B_{\delta',i}(N)\right)$$
$$= P\left((C_{1,i}(\delta) \cap C_{2,i}(\delta) \cap C_{3,i}(\delta))^c \mid B_{\delta',i}(N)\right)$$
$$\leq P\left((C_{1,i}(\delta_{ver}) \cap C_{2,i}(\tfrac{\delta}{2}) \cap C_{3,i}(\min\{\tfrac{\delta_{cont}}{2}, \tfrac{\delta}{2}\}))^c \mid B_{\delta',i}(N)\right)$$
$$\leq P\left((C_{1,i}(\delta_{ver}) \cap \tilde{C}_{2,i}(\delta_{cont}, \tfrac{\delta}{2}))^c \mid B_{\delta',i}(N)\right)$$
$$= P\left(C_{1,i}^c(\delta_{ver}) \cup \tilde{C}_{2,i}^c(\delta_{cont}, \tfrac{\delta}{2}) \mid B_{\delta',i}(N)\right)$$
$$\leq P\left(C_{1,i}^c(\delta_{ver}) \mid B_{\delta',i}(N_0)\right) + P\left(\tilde{C}_{2,i}^c(\delta_{cont}, \tfrac{\delta}{2}) \mid B_{\delta',i}(N)\right)$$
$$< \delta_{ver} + \tfrac{\delta}{2}$$
$$\leq \delta$$

holds for all $N \geq N_0$ by Inclusions (3.19) and (3.20) and Equations (3.16) and (3.18). □

The next lemma is used in the proof of Proposition 3.3.13. It specifies the starting point distances and discretization sizes of the K components in the proof of Proposition 3.3.13 that yield a sequence of approximation errors which results in a total approximation error of at most a beforehand fixed size.

Lemma 3.3.11 *Let $K \geq 1$ and $\delta_{K+1} > 0$. Then there exist pairs $(\delta_1, N_1), \ldots, (\delta_K, N_K)$ with $0 < \delta_1 < \ldots < \delta_{K+1}$ and $N_K \leq \ldots \leq N_1$ such that for all $1 \leq i \leq K$ there exist versions of $(Y_{t,i}^{N,1})_{t \in [0,1]}$ such that*

$$P\left(C_{1,i}\left(\frac{\delta_{i+1}}{2^{K-i+1}}\right) \cap C_{2,i}\left(\frac{\delta_{i+1}}{2^{K-i+1}}\right) \cap C_{3,i}\left(\frac{\delta_{i+1}}{2^{K-i+1}}\right) \mid B_{\delta_i,i}(N)\right) \geq (1 - \delta_{i+1})^{\frac{1}{k}}$$

holds for all $N \geq N_i$. The sets $C_{m,i}$ for $m \in \{1, 2, 3\}$ and $1 \leq i \leq K$ are defined as in Lemma 3.3.10.

Proof. This is a direct consequence of Lemma 3.3.10. Starting with δ_{K+1} Lemma 3.3.10 yields the existence of (δ_K, N_K) with the desired properties which further yields again with Lemma 3.3.10 (δ_{K-1}, N_{K-1}) and so on. After K steps the demanded K pairs are available. □

For $1 \leq i \leq K$ the pair (δ_i, N_i) gives a starting point distance δ_i of the first coordinate of the component processes $(Y_{t,i}^{N,1})_{t \in [0,1]}$ and $(Y_{t,i}^{1})_{t \in [0,1]}$ and associated discretization finenesses $\frac{1}{N_i}$ that ensure a total approximation error for the first coordinate after the i-th step of less than δ_{i+1} ($1 \leq i \leq K$) and especially after the K-th step of at most δ_{K+1} as needed. In addition the second coordinate is controlled by $\frac{\delta_{i+1}}{2^{K-i+1}}$ at each step.

Proof of the Limit Theorem 3.1

After having laid the groundwork before the formulation and proof of the main theorem of this section are given. We show weak convergence of the discrete multiphase processes $(Y_t^N)_{t \in [0,1]}$ to the continuous multiphase process $(Y_t)_{t \in [0,1]}$ introduced in Section 3.3.1 by constructing versions which are close in probability.

Remark 3.3.12 *Since we need a metric on the space of random variables we use one that fits our needs. Therefore we note that*

$$d(f,g) = \sup_{t \in [0,1]} |f_1(t) - g_1(t)| + \lambda^1(\{t \in [0,1] : f_2(t) \neq g_2(t)\})$$

for $f, g \in M$ defines a metric on the function space

$$M = \{f | f : [0,1] \to [0,1] \times \{1,2\} \; t \mapsto (f_1(t), f_2(t)), f_1 \text{ is continuous}, f_2 \text{ is right-continuous}\}.$$

Another useful representation of the second term is given by

$$\lambda^1(\{t \in [0,1] : f_2(t) \neq g_2(t)\}) = \int_0^1 |f_2(t) - g_2(t)| dt$$

since $f_2(t), g_2(t) \in \{1,2\}$ holds for all $t \in [0,1]$.

Theorem 3.1 *Consider the discrete multiphase processes $(Y_t^N)_{t \in [0,1]}$ and the continuous multiphase diffusion process $(Y_t)_{t \in [0,1]}$ on their state spaces $[0,1] \times \{1,2\}$ as introduced in Section 3.3.1. Then the approximation*

$$(Y_t^N)_{t \in [0,1]} \xrightarrow{\mathcal{D}} (Y_t)_{t \in [0,1]}$$

holds for $N \to \infty$.

Proof. By the trivial direction of the Strassen–Dudley Theorem it suffices to show that there exist arbitrarily close versions of $(Y_t^N)_{t \in [0,1]}$ and $(Y_t)_{t \in [0,1]}$. This is the statement of Proposition 3.3.13. □

The following proposition is essential in the proof of Theorem 3.1 and uses all previously derived results.

Proposition 3.3.13 *Consider the processes $(Y_t^N)_{t \in [0,1]}$ and $(Y_t)_{t \in [0,1]}$ as introduced in Section 3.3.1 and let $\delta > 0$. Then there exist versions of $(Y_t^N)_{t \in [0,1]}$ and a $N_1 \geq 1$ such that*

$$P\left(\sup_{t \in [0,1]} |Y_t^{N,1} - Y_t^1| + \lambda^1(\{t \in [0,1] : Y_t^{N,2} \neq Y_t^2\}) > \delta\right) < \delta$$

holds for all $N \geq N_1$. Here λ^1 denotes the one-dimensional Lebesgue measure.

Proof. To prove Proposition 3.3.13 let $\delta > 0$. We show first that it suffices to consider only a finite number of stopping times.

By Proposition 3.3.3 there exists for $\frac{\delta}{2}$ a $K \geq 1$ such that the number of lifetimes, i.e. the number of component processes $(Y_{t,1})_{t\in[0,1]}, (Y_{t,1}^N)_{t\in[0,1]}, (Y_{t,2})_{t\in[0,1]}, (Y_{t,2}^N)_{t\in[0,1]} \ldots$, is stochastically bounded by K, i.e.

$$P\left(\max\left\{\sum_{j=1}^{\infty} 1\!\!1_{\{\sum_{i=1}^{j} \tau_i^N \leq 1\}}, \sum_{j=1}^{\infty} 1\!\!1_{\{\sum_{i=1}^{j} \tau_i \leq 1\}}\right\} > K\right) \leq \frac{\delta}{2} \qquad (3.21)$$

holds independent of the versions of $(Y_t^N)_{t\in[0,1]}$ and for all $N \geq 1$. From now on we consider only the set $A_K = \left\{\max\left\{\sum_{j=1}^{\infty} 1\!\!1_{\{\sum_{i=1}^{j} \tau_i^N \leq 1\}}, \sum_{j=1}^{\infty} 1\!\!1_{\{\sum_{i=1}^{j} \tau_i \leq 1\}}\right\} \leq K\right\}$ and note that $P(A_K) \geq 1 - \frac{\delta}{2}$ holds. Let us recollect the definitions of the considered processes on A_K. From Section 3.3.1 we obtain

$$Y_t^N = \sum_{i=1}^{K} Y_{t-\sum_{j=1}^{i-1}\tau_j^N,i}^N 1\!\!1_{\{\sum_{j=1}^{i-1}\tau_j^N \leq t < \sum_{j=1}^{i}\tau_j^N\}}$$

and

$$Y_t = \sum_{i=1}^{K} Y_{t-\sum_{j=1}^{i-1}\tau_j,i} 1\!\!1_{\{\sum_{j=1}^{i-1}\tau_j \leq t < \sum_{j=1}^{i}\tau_j\}}$$

for $t \in [0,1]$ on A_K.

We see that it suffices to show

$$P\left(A_K \cap \left\{\sup_{t\in[0,1]} |Y_t^{N,1} - Y_t^1| + \lambda^1(\{t \in [0,1] : Y_t^{N,2} \neq Y_t^2\}) > \frac{\delta}{2}\right\}\right) < \frac{\delta}{2}$$

because $P(A_K^c) < \frac{\delta}{2}$. We further obtain the inclusion

$$A_K \cap \left\{\sup_{t\in[0,1]} |Y_t^{N,1} - Y_t^1| + \lambda^1(\{t \in [0,1] : Y_t^{N,2} \neq Y_t^2\}) > \frac{\delta}{2}\right\}$$
$$\subset \left\{\max_{1\leq j\leq K}\left\{\sup_{t\in[\tilde{\tau}_{j-1}^N,\tilde{\tau}_j^N]} |Y_t^{N,1} - Y_t^1|\right\} + \sum_{j=1}^{K} |\tilde{\tau}_j^N - \tilde{\tau}_j| > \frac{\delta}{2}\right\}$$

with $\tilde{\tau}_j^N = \sum_{l=1}^{j} \tau_l^N$ and $\tilde{\tau}_j = \sum_{l=1}^{j} \tau_l$ for $1 \leq j \leq K$.

For $1 \leq j \leq K$ we split the interval $[\tilde{\tau}_{j-1}^N, \tilde{\tau}_j^N]$ disjoint as follows

$$[\tilde{\tau}_{j-1}^N, \tilde{\tau}_j^N] = [\tilde{\tau}_{j-1}^N, \min\{\max\{\tilde{\tau}_{j-1}^N, \tilde{\tau}_{j-1}\}, \tilde{\tau}_j^N\}) \cup [\max\{\tilde{\tau}_{j-1}^N, \tilde{\tau}_{j-1}\}, \min\{\tilde{\tau}_j^N, \tilde{\tau}_j\}]$$
$$\cup (\max\{\min\{\tilde{\tau}_j^N, \tilde{\tau}_j\}, \tilde{\tau}_{j-1}^N\}, \tilde{\tau}_j^N]$$

and note that $[\max\{\tilde{\tau}_{j-1}^N, \tilde{\tau}_{j-1}\}, \min\{\tilde{\tau}_j^N, \tilde{\tau}_j\}] = [\tilde{\tau}_{j-1}^N, \tilde{\tau}_j^N] \cap [\tilde{\tau}_{j-1}, \tilde{\tau}_j]$. Denote the sets above by $T_{j,l}$ for $l \in \{1, 2, 3\}$ as follows

$$\begin{aligned}
T_{j,1} &= [\tilde{\tau}_{j-1}^N, \min\{\max\{\tilde{\tau}_{j-1}^N, \tilde{\tau}_{j-1}\}, \tilde{\tau}_j^N\}) = [\tilde{\tau}_{j-1}^N, \min\{\tilde{\tau}_{j-1}, \tilde{\tau}_j^N\}), \\
T_{j,2} &= [\max\{\tilde{\tau}_{j-1}^N, \tilde{\tau}_{j-1}\}, \min\{\tilde{\tau}_j^N, \tilde{\tau}_j\}], \\
T_{j,3} &= (\max\{\min\{\tilde{\tau}_j^N, \tilde{\tau}_j\}, \tilde{\tau}_{j-1}^N\}, \tilde{\tau}_j^N] = (\max\{\tilde{\tau}_j, \tilde{\tau}_{j-1}^N\}, \tilde{\tau}_j^N].
\end{aligned}$$

A sketch of the situation is given in Figure 3.5.

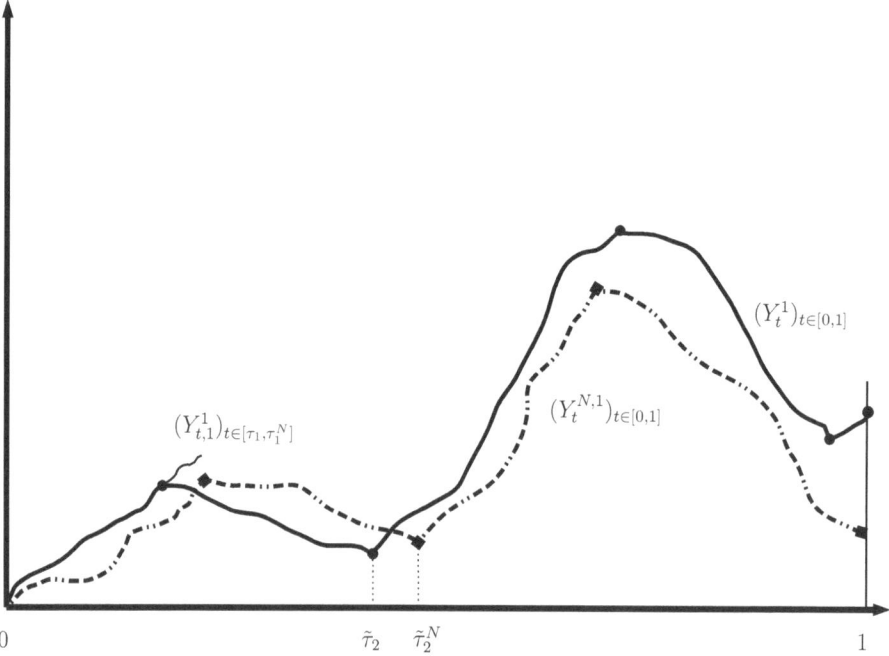

Figure 3.5: Sketch of the first coordinates of the continuous (solid line, circles) and the discrete (dashed line, squares) processes. The continuous process has 5 stopping times marked by circles and the discrete process has 4 stopping times marked by squares. The continuation of the first coordinate of the first component of the continuous process is shown by a small solid line.

We estimate

$$\left|Y_t^{N,1} - Y_t^1\right| = \left|Y_t^{N,1} - Y_{t-\tilde{\tau}_{j-1}^N,j}^1 + Y_{t-\tilde{\tau}_{j-1}^N,j}^1 - Y_{\tilde{\tau}_{j-1}}^1 + Y_{\tilde{\tau}_{j-1}}^1 - Y_t^1\right|$$

$$\leq \left|Y_{t-\tilde{\tau}_{j-1}^N,j}^{N,1} - Y_{t-\tilde{\tau}_{j-1}^N,j}^1\right| + \left|Y_{t-\tilde{\tau}_{j-1}^N,j}^1 - Y_{0,j}^1\right| + \left|Y_{\tilde{\tau}_{j-1}}^1 - Y_t^1\right|$$

because $Y_{0,j}^1 = Y_{\tilde{\tau}_{j-1}}^1$. Hence we obtain on $T_{j,1}$

$$\sup_{t\in T_{j,1}} \left|Y_t^{N,1} - Y_t^1\right| \leq \sup_{t\in T_{j,1}} \left|Y_{t-\tilde{\tau}_{j-1}^N,j}^{N,1} - Y_{t-\tilde{\tau}_{j-1}^N,j}^1\right| + \sup_{t\in T_{j,1}} \left|Y_{t-\tilde{\tau}_{j-1}^N,j}^1 - Y_{0,j}^1\right|$$

$$+ \sup_{t\in T_{j,1}} \left|Y_{\tilde{\tau}_{j-1}}^1 - Y_t^1\right|$$

$$\leq \sup_{t\in[0,\tilde{\tau}_{j-1}-\tilde{\tau}_{j-1}^N)} \left|Y_{t,j}^{N,1} - Y_{t,j}^1\right| + \sup_{t\in[0,\tilde{\tau}_{j-1}-\tilde{\tau}_{j-1}^N)} \left|Y_{t,j}^1 - Y_{0,j}^1\right|$$

$$+ \sup_{t\in[\tilde{\tau}_{j-1}^N,\tilde{\tau}_{j-1})} \left|Y_{\tilde{\tau}_{j-1}}^1 - Y_t^1\right|$$

$$\leq \sup_{t\in[0,1]} \left|Y_{t,j}^{N,1} - Y_{t,j}^1\right| + \sup_{|t-s|<|\tilde{\tau}_{j-1}-\tilde{\tau}_{j-1}^N|} \left|Y_{t,j}^1 - Y_{s,j}^1\right|$$

$$+ \sup_{|t-s|<|\tilde{\tau}_{j-1}-\tilde{\tau}_{j-1}^N|} \left|Y_s^1 - Y_t^1\right|$$

for $1 \leq j \leq K$. Analogous considerations on $T_{j,3}$ yield

$$\sup_{t\in T_{j,3}} \left|Y_t^{N,1} - Y_t^1\right| \leq \sup_{t\in T_{j,3}} \left|Y_{t-\tilde{\tau}_{j-1}^N,j}^{N,1} - Y_{t-\tilde{\tau}_{j-1}^N,j}^1\right| + \sup_{t\in T_{j,3}} \left|Y_{t-\tilde{\tau}_{j-1}^N,j}^1 - Y_{\tau_j,j}^1\right|$$

$$+ \sup_{t\in T_{j,1}} \left|Y_{\tilde{\tau}_j}^1 - Y_t^1\right|$$

$$\leq \sup_{t\in[0,\tau_j^N)} \left|Y_{t,j}^{N,1} - Y_{t,j}^1\right| + \sup_{t\in[\max\{0,\tilde{\tau}_j-\tilde{\tau}_{j-1}^N\},\tau_j^N)} \left|Y_{t,j}^1 - Y_{\tau_j,j}^1\right|$$

$$+ \sup_{t\in[\tilde{\tau}_j,\tilde{\tau}_j^N)} \left|Y_{\tilde{\tau}_j}^1 - Y_t^1\right|$$

$$\leq \sup_{t\in[0,1]} \left|Y_{t,j}^{N,1} - Y_{t,j}^1\right| + \sup_{|t-s|<\max\{|\tilde{\tau}_{j-1}-\tilde{\tau}_{j-1}^N|,|\tau_j^N-\tau_j|\}} \left|Y_{t,j}^1 - Y_{s,j}^1\right|$$

$$+ \sup_{|t-s|<|\tilde{\tau}_j-\tilde{\tau}_j^N|} \left|Y_s^1 - Y_t^1\right|.$$

for $1 \leq j \leq K$. For $T_{j,2}$

$$\sup_{t\in T_{j,2}} \left|Y_t^{N,1} - Y_t^1\right| = \sup_{t\in T_{j,2}} \left|Y_{t-\tilde{\tau}_{j-1}^N,j}^{N,1} - Y_{t-\tilde{\tau}_{j-1},j}^1\right|$$

$$\leq \sup_{t\in T_{j,2}} \left|Y_{t-\tilde{\tau}_{j-1}^N,j}^{N,1} - Y_{t-\tilde{\tau}_{j-1}^N,j}^1\right| + \sup_{t\in T_{j,2}} \left|Y_{t-\tilde{\tau}_{j-1}^N,j}^1 - Y_{t-\tilde{\tau}_{j-1},j}^1\right|$$

$$\leq \sup_{t\in[0,1]} \left|Y_{t,j}^{N,1} - Y_{t,j}^1\right| + \sup_{|t-s|\leq|\tilde{\tau}_{j-1}^N-\tilde{\tau}_{j-1}|} \left|Y_{t,j}^1 - Y_{s,j}^1\right|$$

holds for $1 \leq j \leq K$. Altogether we obtain

$$\sup_{t \in [\tilde{\tau}_{j-1}^N, \tilde{\tau}_j^N]} |Y_t^{N,1} - Y_t^1| \leq \sup_{t \in [0,1]} \left| Y_{t,j}^{N,1} - Y_{t,j}^1 \right|$$
$$+ \sup_{|t-s| < \max\{|\tilde{\tau}_{j-1} - \tilde{\tau}_{j-1}^N|, |\tau_j^N - \tau_j|\}} \left| Y_{t,j}^1 - Y_{s,j}^1 \right| \quad (3.22)$$
$$+ \sup_{|t-s| < \max\{|\tilde{\tau}_{j-1} - \tilde{\tau}_{j-1}^N|, |\tilde{\tau}_j - \tilde{\tau}_j^N|\}} |Y_s^1 - Y_t^1|$$

for $1 \leq j \leq K$.

Now we invoke Lemma 3.3.11 with $\delta_{K+1} = \frac{\delta}{4}$, fix $N \geq N_1$ and consider the set

$$\hat{C} = \bigcap_{1 \leq j \leq K} \left(C_{1,j}\left(\frac{\delta_{j+1}}{2^{K-j+1}}\right) \cap C_{2,j}\left(\frac{\delta_{j+1}}{2^{K-j+1}}\right) \cap C_{3,j}\left(\frac{\delta_{j+1}}{2^{K-j+1}}\right) \right)$$

where the sets $C_{m,j}(\delta)$ are defined in Lemma 3.3.10 for $1 \leq j \leq K$, $m \in \{1,2,3\}$ and $\delta > 0$. On $\hat{C}$

$$\sum_{l=1}^{j} |\tau_l^N - \tau_l| < \sum_{l=1}^{j} \frac{\delta_{l+1}}{2^{K-l+1}} \leq \frac{\delta_{j+1}}{2^{K-j}} \leq \delta_{K+1} \leq \frac{\delta}{4}$$

holds for $1 \leq j \leq K$. Further we see that the two estimates

$$\max\{|\tilde{\tau}_{j-1} - \tilde{\tau}_{j-1}^N|, |\tau_j^N - \tau_j|\} \leq \max\{\sum_{l=1}^{j-1}|\tau_l^N - \tau_l|, |\tau_j^N - \tau_j|\} \leq |\tau_j^N - \tau_j| + \frac{\delta_j}{2^{K-j+1}},$$
$$\max\{|\tilde{\tau}_{j-1} - \tilde{\tau}_{j-1}^N|, |\tilde{\tau}_j - \tilde{\tau}_j^N|\} \leq |\tau_j^N - \tau_j| + \sum_{l=1}^{j-1}|\tau_l^N - \tau_l| \leq |\tau_j^N - \tau_j| + \frac{\delta_j}{2^{K-j+1}}$$

hold for $1 \leq j \leq K$ on $\hat{C}$. This yields with Estimate (3.22) and the definition of $\hat{C}$

$$\hat{C} \subset \left\{ \max_{1 \leq j \leq K} \left\{ \sup_{t \in [\tilde{\tau}_{j-1}^N, \tilde{\tau}_j^N]} |Y_t^{N,1} - Y_t^1| \right\} + \sum_{j=1}^K |\tau_j^N - \tau_j| \leq \frac{\delta}{2} \right\}.$$

We define

$$C_j = C_{1,j}\left(\frac{\delta_{j+1}}{2^{K-j+1}}\right) \cap C_{2,j}\left(\frac{\delta_{j+1}}{2^{K-j+1}}\right) \cap C_{3,j}\left(\frac{\delta_{j+1}}{2^{K-j+1}}\right)$$

for $1 \leq j \leq K$ and apply the multiplication formula to obtain

$$P(\hat{C}) = P\left(\bigcap_{1 \leq j \leq K} C_j\right) = \prod_{j=1}^K P\left(C_j \mid \bigcap_{1 \leq l \leq j-1} C_l\right).$$

A useful relation is given by

$$P\left(C_j \mid \bigcap_{1 \leq l \leq j-1} C_l\right) \geq P(C_j \mid B_{\delta_j, j}(N))$$

for $1 \leq i \leq K$. By Lemma 3.3.11 the chosen versions of $(Y_{t,i}^{N,1})_{t \in [0,1]}$ suffice

$$P(C_j \mid B_{\delta_j,j}(N)) \geq (1 - \delta_{j+1})^{\frac{1}{K}} \geq (1 - \delta_{K+1})^{\frac{1}{K}} \geq \left(1 - \frac{\delta}{4}\right)^{\frac{1}{K}}$$

for $1 \leq j \leq K$ and $N \geq N_j$. This implies

$$P\left(\max_{1 \leq j \leq K} \left\{\sup_{t \in [\tilde{\tau}_{j-1}^N, \tilde{\tau}_j^N]} |Y_t^{N,1} - Y_t^1|\right\} + \sum_{j=1}^K |\tau_j^N - \tau_j| \leq \frac{\delta}{2}\right) \geq P(\hat{C}) \geq 1 - \frac{\delta}{4}$$

for $N \geq N_1$ to finish the proof. $\square$

Remark 3.3.14 *In Section 3.3 an approximation theorem for a twophase process on $[0,1] \times \{1,2\}$ by an associated sequence of Markov birth–death chains is stated and proved. We like to stress that this result extends without much further difficulty to multiphase processes on $[0,T] \times \{1,2,\ldots,K\}$ for arbitrary $T > 0$. Only the desire to give a clear exposition containing all essential ideas and techniques led to the restriction to twophase processes on $[0,1] \times \{1,2\}$. The extension of the state space $[0,1] \times \{1,2\}$ to $[0,T] \times \{1,2\}$ is quite obvious since the presented proofs all hold for spaces of the form $[0,T] \times \{1,2\}$. The second extension from $[0,T] \times \{1,2\}$ to $[0,T] \times \{1,2,\ldots,K\}$ needs a stronger effort since a random variable determining the target states in the second coordinate after a phase change has occured has to be introduced and considered. But similar and even simpler techniques than the ones presented above are sufficient, especially the coupling of those new random variables of the discrete and the continuous processes analogous to the one used for the stopping times.*

3.4 Partial Differential Equation for the Diffusion Limit

In the previous section it is proved that the discrete Markov chain models and their corresponding multiphase models have continuous analogues which can be realized as limits of discrete Markov chains. These continuous Markov processes are often governed by certain differential equations, e.g. as shown in Section 2.3. In the present section we derive such a differential equation for a multiphase process using the approximation scheme from Section 2.3.1. We consider a birth–death Markov chain $(X_n)_{n\geq 0}$ on a state space

$$S = \{1, 2, \ldots, N\} \times \{1, 2, \ldots, K\}$$

with $N, K \in \mathbb{N} \setminus \{0\}$ and transition probabilities

$$\begin{aligned}
p_{(i,k)(i-1,k)} &= \delta_i^{(k)}, \\
p_{(i,k)(i+1,k)} &= \beta_i^{(k)}, \\
p_{(i,k)(i,l)} &= \lambda_i^{(kl)}
\end{aligned} \tag{3.23}$$

in the interior, i.e. $2 \leq i \leq N-1$ and $1 \leq l, k \leq K$, sufficing

$$\delta_i^{(k)} + \beta_i^{(k)} + \sum_{l=1}^{K} \lambda_i^{(kl)} = 1 \tag{3.24}$$

for all $(i, k) \in S$. Boundary conditions are defined later and examined separately.

We use the Markov chain $(X_n)_{n\geq 0} = (Y_n, Z_n)_{n\geq 0}$ where $Y_n : \Omega \to \{1, 2, \ldots, N\}$ and $Z_n : \Omega \to \{1, 2, \ldots, K\}$ are the projections of X_n on the first respectively second coordinate to define a new one on $[0, 1] \times \{1, 2, \ldots, K\}$ by setting

$$X_t^\Delta = \left(\Delta Y_{[\frac{t}{\varepsilon}]}, Z_{[\frac{t}{\varepsilon}]} \right)$$

for all $t \geq 0$ with the discretization parameters chosen as $\Delta = \frac{1}{N}$ and $\varepsilon = \frac{\Delta^2}{\max_{x \in [0,1], 1 \leq k \leq K} D_k(x)}$ as in Section 2.3.1 and transition probabilities as in Equations (3.23) with diffusion functions D_k as introduced below.

Let now $D_k : [0, 1] \to (0, \infty)$, $v_k : [0, 1] \to \mathbb{R}$ and $\lambda_{kl} : [0, 1] \to [0, \infty)$ be continuous diffusion, drift and phase change functions for all $1 \leq k, l \leq K$ such that the transition probabilities for X_t^Δ suffice Equation (3.24) and equations

$$\begin{aligned}
\delta_i^{(k)} &= \tfrac{\varepsilon}{2\Delta^2} D_k(i\Delta) - \tfrac{\varepsilon}{2\Delta} v_k(i\Delta), \\
\beta_i^{(k)} &= \tfrac{\varepsilon}{2\Delta^2} D_k(i\Delta) + \tfrac{\varepsilon}{2\Delta} v_k(i\Delta), \\
\lambda_i^{(kl)} &= \varepsilon \lambda_{kl}(i\Delta),
\end{aligned} \tag{3.25}$$

for all $1 \leq i \leq N$ and $1 \leq k, l \leq K$.

3.4.1 Interior

We start by deriving a set of partial differential equations for the interior $(0,1)$ of the interval $[0,1]$. We show that there exists one partial differential equation for each state k in the second phase $\{1, 2, \ldots, K\}$. Under consideration is only a continuous first coordinate although a continuous second coordinate is imaginable as well. Similar techniques as the ones applied in this section would be sufficient to deduce analogous results in such a case.

Denote by $p_\Delta^{(k)}(n\varepsilon, i\Delta)$ the probability that the chain X_t^Δ is in state $(i\Delta, k)$ at time $n\varepsilon$, i.e. $p_\Delta^{(k)}(n\varepsilon, i\Delta) = P(X_{n\varepsilon}^\Delta = (i\Delta, k))$ and obtain the identity

$$p_\Delta^{(k)}((n+1)\varepsilon, i\Delta) = \delta_{i+1}^{(k)} p_\Delta^{(k)}(n\varepsilon, (i+1)\Delta) + \beta_{i-1}^{(k)} p_\Delta^{(k)}(n\varepsilon, (i-1)\Delta)$$
$$+ \sum_{l=1}^{N_2} \lambda_i^{(lk)} p_\Delta^{(l)}(n\varepsilon, i\Delta)$$

for all $n \geq 0$, $1 \leq k \leq K$ and $2 \leq i \leq N-1$. This yields with Equation (3.25)

$$p_\Delta^{(k)}((n+1)\varepsilon, i\Delta) - p_\Delta^{(k)}(n\varepsilon, i\Delta)$$
$$= \delta_{i+1}^{(k)} p_\Delta^{(k)}(n\varepsilon, (i+1)\Delta) + \beta_{i-1}^{(k)} p_\Delta^{(k)}(n\varepsilon, (i-1)\Delta)$$
$$- [\beta_i^{(k)} + \delta_i^{(k)} + \sum_{l=1, l \neq k}^{K} \lambda_i^{(kl)}] p_\Delta^{(k)}(n\varepsilon, i\Delta) + \sum_{l=1, l \neq k}^{K} \lambda_i^{(lk)} p_\Delta^{(l)}(n\varepsilon, i\Delta)$$
$$= (\tfrac{\varepsilon}{2\Delta^2} D_k((i+1)\Delta) - \tfrac{\varepsilon}{2\Delta} v_k((i+1)\Delta)) p_\Delta^{(k)}(n\varepsilon, (i+1)\Delta)$$
$$+ (\tfrac{\varepsilon}{2\Delta^2} D_k((i-1)\Delta) + \tfrac{\varepsilon}{2\Delta} v_k((i-1)\Delta)) p_\Delta^{(k)}(n\varepsilon, (i-1)\Delta)$$
$$- [\tfrac{\varepsilon}{\Delta^2} D_k(i\Delta) + \sum_{l=1, l \neq k}^{K} \varepsilon \lambda_{kl}(i\Delta)] p_\Delta^{(k)}(n\varepsilon, i\Delta) + \sum_{l=1, l \neq k}^{K} \varepsilon \lambda_{lk}(i\Delta) p_\Delta^{(l)}(n\varepsilon, i\Delta)$$

for all $n \geq 0$, $1 \leq k \leq K$ and $2 \leq i \leq N-1$. It can further be rearranged to

$$\frac{p_\Delta^{(k)}((n+1)\varepsilon, i\Delta) - p_\Delta^{(k)}(n\varepsilon, i\Delta)}{\varepsilon} = \frac{D_k((i+1)\Delta) p_\Delta^{(k)}(n\varepsilon, (i+1)\Delta) - 2 D_k(i\Delta) p_\Delta^{(k)}(n\varepsilon, i\Delta) + D_k((i-1)\Delta) p_\Delta^{(k)}(n\varepsilon, (i-1)\Delta)}{2\Delta^2}$$
$$- \frac{v_k((i+1)\Delta) p_\Delta^{(k)}(n\varepsilon, (i+1)\Delta) - v_k((i-1)\Delta) p_\Delta^{(k)}(n\varepsilon, (i-1)\Delta)}{2\Delta}$$
$$+ \sum_{l=1, l \neq k}^{K} \lambda_{lk}(i\Delta) p_\Delta^{(l)}(n\varepsilon, i\Delta)$$
$$- \sum_{l=1, l \neq k}^{K} \lambda_{kl}(i\Delta) p_\Delta^{(k)}(n\varepsilon, i\Delta)$$

for all $n \geq 0$, $1 \leq k \leq K$ and $2 \leq i \leq N-1$. Then the equation above tends to

$$\begin{aligned}\frac{\partial}{\partial t}p_k(t,x) &= \tfrac{1}{2}\tfrac{\partial^2}{\partial x^2}\left[D_k(x)p_k(t,x)\right] - \tfrac{\partial}{\partial x}\left[v_k(x)p_k(t,x)\right] \\ &+ \sum_{l=1,l\neq k}^{K}[\lambda_{lk}(x)p_l(t,x) - \lambda_{kl}(x)p_k(t,x)]\end{aligned} \quad (3.26)$$

for all $t \geq 0$, $x \in (0,1)$ and $1 \leq k \leq K$ if $p_\Delta^{(k)}(n\varepsilon, i\Delta)$ converges and the relations $n\varepsilon \approx t$ and $i\Delta \approx x$ hold.

We see that Equation (3.26) is a version of the Fokker–Planck equation for the density $p(t,x)$ with additional terms corresponding to the phase transitions. With an approach as in Section 2.3.4 the corresponding backward equation would have been derived.

3.4.2 Boundaries

After deriving a set of partial differential equations for the interior $(0,1)$ we examine the boundaries. We consider a reflecting boundary to the left, i.e. at $x = 0$ respectively state $i = 1$ and an absorbing boundary to the right, i.e. at $x = 1$ respectively state $i = N$. As for the interior again a set of K boundary conditions for each boundary corresponding to the states in the second phase is discovered. Surprisingly the phase changes do not appear in the boundary conditions although they are part of the partial differential equations in the interior. The reason for this is found in the fact that when a phase change occurs the first coordinate of the process is unaffected.

Reflecting Boundary

Setting $\delta_1^{(k)} = 0$ corresponds to an reflecting boundary. The same approach as above for the interior yields

$$p_\Delta^{(k)}((n+1)\varepsilon, \Delta) = \delta_2^{(k)}p_\Delta^{(k)}(n\varepsilon, 2\Delta) + \sum_{l=1}^{K}\lambda_1^{(lk)}p_\Delta^{(l)}(n\varepsilon, \Delta)$$

for all $n \geq 0$ and $1 \leq k \leq K$. Using Equation (3.25) we obtain

$$\begin{aligned}
p_\Delta^{(k)}((n+1)\varepsilon, \Delta) - p_\Delta^{(k)}(n\varepsilon, \Delta) &= \delta_2^{(k)} p_\Delta^{(k)}(n\varepsilon, 2\Delta) - [\beta_1^{(k)} + \textstyle\sum_{l=1,l\neq k}^{K} \lambda_1^{(kl)}] p_\Delta^{(k)}(n\varepsilon, \Delta) \\
&\quad + \textstyle\sum_{l=1,l\neq k}^{K} \lambda_i^{(lk)} p_\Delta^{(l)}(n\varepsilon, \Delta) \\
&= (\tfrac{\varepsilon}{2\Delta^2} D_k(2\Delta) - \tfrac{\varepsilon}{2\Delta} v_k(2\Delta)) p_\Delta^{(k)}(n\varepsilon, 2\Delta) \\
&\quad - [\tfrac{\varepsilon}{2\Delta^2} D_k(\Delta) + \tfrac{\varepsilon}{2\Delta} v_k(\Delta) + \textstyle\sum_{l=1,l\neq k}^{K} \varepsilon \lambda_{kl}(\Delta)] p_\Delta^{(k)}(n\varepsilon, \Delta) \\
&\quad + \textstyle\sum_{l=1,l\neq k}^{K} \varepsilon \lambda_{lk}(\Delta) p_\Delta^{(l)}(n\varepsilon, \Delta)
\end{aligned}$$

for all $n \geq 0$ and $1 \leq k \leq K$. Further transformation results in

$$\begin{aligned}
\Delta \tfrac{p_\Delta^{(k)}((n+1)\varepsilon, \Delta) - p_\Delta^{(k)}(n\varepsilon, \Delta)}{\varepsilon} &= \tfrac{D_k(2\Delta) p_\Delta^{(k)}(n\varepsilon, 2\Delta) - D_k(\Delta) p_\Delta^{(k)}(n\varepsilon, \Delta)}{2\Delta} \\
&\quad - \tfrac{1}{2}(v_k(\Delta) p_\Delta^{(k)}(n\varepsilon, \Delta) + v_k(2\Delta) p_\Delta^{(k)}(n\varepsilon, 2\Delta)) \\
&\quad + \Delta \textstyle\sum_{l=1,l\neq k}^{K}(\lambda_{lk}(\Delta) p_\Delta^{(l)}(n\varepsilon, \Delta) - \lambda_{kl}(\Delta) p_\Delta^{(k)}(n\varepsilon, \Delta))
\end{aligned}$$

for all $n \geq 0$ and $1 \leq k \leq K$. The limit in this case under the same assumptions as for Equation (3.26) looks as follows

$$0 = \frac{1}{2} \frac{\partial}{\partial x} [D_k(x) p_k(t, x)] |_{x=0} - v_k(0) p_k(t, 0) \tag{3.27}$$

for all $t \geq 0$.

Absorbing Boundary

Letting $\lambda_N^{kk} = 1$ for all $1 \leq k \leq K$ creates an absorbing boundary. In this case we find that

$$p_\Delta^{(k)}((n+1)\varepsilon, N\Delta) = \beta_{N-1}^{(k)} p_\Delta^{(k)}(n\varepsilon, (N-1)\Delta) + p_\Delta^{(k)}(n\varepsilon, N\Delta)$$

holds for all $n \geq 0$ and $1 \leq k \leq K$. This leads again with Equation (3.25) to

$$\Delta^2 \frac{p_\Delta^{(k)}((n+1)\varepsilon, N\Delta) - p_\Delta^{(k)}(n\varepsilon, N\Delta)}{\varepsilon} = \frac{1}{2}(D_k((N-1)\Delta) + \Delta v_k((N-1)\Delta)) p_\Delta^{(k)}(n\varepsilon, (N-1)\Delta)$$

which is valid for all $n \geq 0$ and $1 \leq k \leq K$. Assumptions as for Equation (3.26) give the limit

$$0 = \lim_{x \nearrow 1} p_k(t, x) \qquad (3.28)$$

for all $t \geq 0$ and $1 \leq k \leq K$.

3.4.3 Results

In the previous part of this section sets of partial differential equations with boundary conditions for reflecting and absorbing boundaries which belong to a continuous analogue via a birth–death approximation with additional phase transitions of the presented multiphase process are derived. We collect our results in the following theorem.

Theorem 3.2 *Under convergence assumptions the density $p(t, x)$ of the continuous Markov process corresponding to the multiphase process with an reflecting and an absorbing boundary introduced in Chapter 3 suffices the set of partial differential equations from Equations (3.26) with boundary conditions from Equations (3.27) and (3.28)*

$$\frac{\partial}{\partial t} p_k(t, x) = \frac{1}{2} \frac{\partial^2}{\partial x^2} [D_k(x) p_k(t, x)] - \frac{\partial}{\partial x} [v_k(x) p_k(t, x)]$$
$$+ \sum_{l=1, l \neq k}^{K} [\lambda_{lk}(x) p_l(t, x) - \lambda_{kl}(x) p_k(t, x)],$$
$$0 = \frac{1}{2} \frac{\partial}{\partial x} [D_k(x) p_k(t, x)]|_{x=0} - v_k(0) p_k(t, 0),$$
$$0 = \lim_{x \nearrow 1} p_k(t, x),$$

for all $t \geq 0$, $x \in (0, 1)$ and $1 \leq k \leq K$.

Proof. The proof of this theorem is content of the current Section above Theorem 3.2. □

Chapter 4

Slugging Fluidized Beds

So far chiefly homogeneous Markov chain models have been investigated. In opposite to this the current chapter places a heterogeneous Markov chain model in the center of investigation. Related to the bubbling fluidized bed reactors from examples in Sections 2.2.4 and 3.1 are slugging fluidized beds. These beds contain larger bubbles socalled slugs and show a quite different behavior than bubbling fluidized beds. They constitue a complex system that is not easily modeled deterministically. A successful approach of inventing a Markov chain model for such a slugging fluidized bed is presented in the following which is based on the previous work of Dehling, Dechsiri, Gottschalk, Wright and Hoffmann (2006).

The chapter is divided into three parts each containing different aspects. The first part presents a brief description of slugging fluidized beds as a phenomenon in physical and engineering science. This is followed by a section where the stochastic model is presented. The last section of this chapter discusses results of application of the before presented model via comparing results derived from the model with two sets of experimental data.

4.1 Slugging Fluidized Beds

In the same setting as for bubbling fluidized beds as in examples in Sections 2.2.4 and 3.1 slugging fluidization can occur if the bed has a high aspect ratio. Again gas is injected in a reactor through a porous distributor plate at the bottom such that at a certain velocity of the gas the powder inside the reactor starts to float and exhibit a fluid-like behavior. This causes fluidization in the bed disturbed by bubbles forming above the distributor plate and rising to the top of the reactor. The particular feature of slugging fluidized beds shows itself in the nature of the bubbles. These have a diameter of almost the width of the reactor and are then called slugs. Figure 4.1 shows two types of slugs.

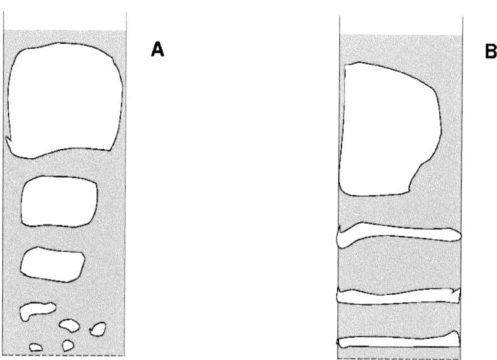

Figure 4.1: Modes of Slugging: **A** axisymmetric and **B** wall slugs (Baeyens and Geldart, 1974).

The behavior of particles inside a slugging fluidized bed reactor differs from those in a bubbling bed reactor. The slugs form at the bottom and push particles above them upwards. Only a small amount of particles above a slug moves downwards through the annulus between reactor and slug. This causes a mixing of particles in the reactor but a different one than in bubbling fluidized bed reactors. In contrast to bubbling fluidized bed reactors which are discussed in examples in Sections 2.2.4 and 3.1 when operated continuously, i.e. with some in- and some outflow we discuss closed slugging fluidized reactors without any in- or outflow. For more background on slugging fluidized beds we refer to Davidson and Harrison (1971), Matsen et al. (1969) and Schouten et al. (1988).

4.2 A Markov Chain Model for a Slugging Fluidized Bed

To present the Markov chain model for a slugging fluidized bed we choose to start with a discussion of the model setup. This includes the breaking up of the process into several aspects which are separately investigated. These are slug formation, slug rise, segregation and continuity conditions. Afterwards the Markov chain is introduced containing all the characteristic aspects discussed before whereupon we elaborate separately.

4.2.1 Model Setup

We consider a slugging fluidized bed with a binary mixture of particles which causes a segregation effect for the particles. The particles of higher density tend to sink downwards while the particles of lower density are more frequently found in the upper parts of the reactor. The movement of a particle inside the reactor during some time period is governed by the occurences of consecutive slugs during this period. We envision the existence of two separate phases which describe the occurence of one slug. First the slug forms at some height in the reactor and grows in size until it detaches and moves upwards. When it hits the top of the bed it collapses and causes the bed height to revert back to its original height. After describing one slug and its effects the effect of several slugs is found via superposition.

Slug Formation

The slug forms at some height h_0 in the bed and grows in size with some constant velocity v until it reaches a size or height h_s and detaches to move upward. While the slug forms all particles in the bed are pushed upwards and the bed expands from height h_b to height $H = h_b + h_s$. A sketch of this can be seen in Figure 4.2 in the middle.

Slug Rise

When the slug has grown to its eventual size h_s it detaches and moves upwards while keeping its velocity v. The heights of the bed H and the slug h_s remain constant throughout this phase. Hitting the top of the bed finally kills the slug, it collapses and the bed height reverts back to h_b.

In the slug rise phase the particles are not exclusively pushed upwards but they can also possibly stay at their current height when they are still above the slug or move downwards below the slug through the annulus. We sketch this situation on the right in Figure 4.2.

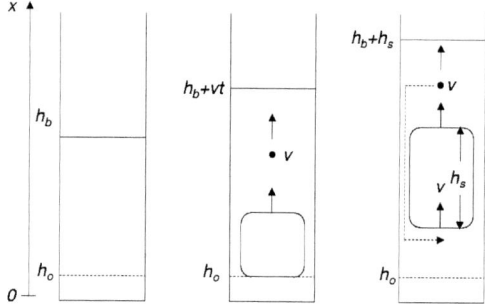

Figure 4.2: Slugging fluidized bed in different states of development: **A** Bed at rest, **B** Slug formation phase and **C** Slug rise phase. (The dot represents a single particle)

The probability density for a particle to move downwards through the annulus is assumed to be a decreasing function $\lambda_h : [h, H] \to [0, \infty)$ of the vertical distance from the slug nose height h to the particle as pictured in Figure 4.3.

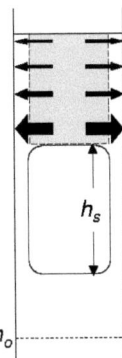

Figure 4.3: A sketch indicating the particle movement to the annulus in front of slug nose. Most of the particles that deposit underneath the slug come from the region close to the slug nose rather than from higher in the bed. The thickness of arrows indicate the magnitude of the particle flux

Segregation

Since we consider a binary mixture of different particles we have to model the segregation effect that takes place as well. The denser particles are more likely to move downwards than their less dense counterparts. This is done later by modifying the transition probabilities accordingly.

Continuity Condition

The mass in the bed stays constant over time and the part of the reactor occupied by the slug cannot contain particles. Therefore a mass balance equation has to hold which assures that the flow downwards through the annulus balances the upwards movement of the slug. This leads to the following continuity condition

$$\int_{h}^{H} \lambda_h(x) dx = v, \tag{4.1}$$

valid for all $h \in (h_s, H)$. The particles above the slug that do not flow downwards either stay at their current height or move upwards with some velocity $v_h : [h, H] \to (0, h_s]$ depending on the height of the slug nose $h \in [h_b, H]$ and sufficing the continuity equation

$$v_h(x) = v - \int_h^x \lambda_h(y) dy$$

for all $x \in (h, H)$. With Equation (4.1) we obtain the equation

$$v_h(x) = \int_x^H \lambda_h(y) dy \qquad (4.2)$$

for all $x \in (h, H)$.

4.2.2 The Markov Chain Model

Now that the model is set up we can formulate a Markov chain that exhibits its features and characteristics. We discretize the bed into cells of height $\Delta > 0$ and time into steps $\varepsilon > 0$ such that the slug grows or rises one cell per timestep, i.e. $\frac{\Delta}{\varepsilon} = v$.

For simplicity of the presentation we assume $h_0 = 0$ and the heights of the slug and the bed h_s and h_b to be integer multipliers of Δ, i.e. there exist $i_s, i_b \in \mathbb{N}$ such that $h_s = i_s \Delta$ and $h_b = i_b \Delta$ hold. We number the cells from 1 to $N = i_s + i_b$ and consider a Markov chain $(X_n)_{n \geq 0}$ on $\{1, 2, \ldots, N\}$ where X_n describes the height of a particle in the bed at time n. The starting distribution is given by $\pi = (\pi_i)_{1 \leq i \leq N}$, i.e. $P(X_0 = i) = \pi_i$ holds for all $i \in \{1, 2, \ldots, N\}$ and the transition probabilities are defined below. Further the starting distribution is required to suffice $\pi_i = 0$ for all $i \in \{i_b + 1, \ldots, N\}$ since the upper part of the bed does not even exist at time 0. We sketch the Markov chain in Figure 4.4 below.

Slug Formation

The slug formation is essentially a deterministic process. In timesteps 1 to i_s the slug forms and all particles are pushed upwards by one cell at each timestep, i.e.

$$P(X_n = i+1 | X_{n-1} = i) = p(n)_{i,i+1} = 1$$

holds for all $n \in \{1, 2, \ldots, i_s\}$ and $i \in \{n, \ldots, i_b + n\}$. For sake of completeness we set $p(n)_{i,i} = 1$ for all $n \in \{1, 2, \ldots, i_s\}$ and $i \in \{1, \ldots, n-1\} \cup \{i_b+n+1, \ldots, N\}$. To simplify this phase we assume that the slug has already formed and shift the starting distribution π with the matrix $S = (s_{ij})_{1 \leq i,j \leq N}$ defined by

$$s_{ij} := \begin{cases} 1 & if\ 1 \leq i \leq i_b\ and\ j = i + i_s, \\ 0 & else, \end{cases} \qquad (4.3)$$

to $\bar{\pi} = \pi S$. Hence we start counting time just after the slug has formed, such that the slug already rises in the timestep from time 0 to time 1.

Slug Rise

After shifting the starting distribution the slug rises and the process loses its deterministic nature. Now three different transitions for particles above the slug, i.e. in states i with $i \in \{i_s + 1 + n, \ldots, N\}$ at time $n \in \{1, \ldots, i_b\}$ to some state $j \in \{i, i+1, n\}$ are possible.

- Move to what will be the first cell below the slug after the time step, i.e. to state $j = n$.
- Move one cell up, i.e. to $j = i + 1$.
- Stay in the same cell, i.e. $j = i$.

A sketch is given with Figure 4.4.

These transitions take place between time 1 and time i_b because at time i_b the slug reaches the top and collapses. Afterwards the process starts anew.

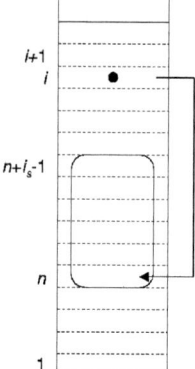

Figure 4.4: The Markov chain model for the n-th transition. Just before the transition the slug occupies the cells $n, \ldots, n + i_s - 1$. The arrow indicates the transition into the annulus, i.e. from cell i to cell n. The two other possible transitions are up into cell $i + 1$ and staying in cell i.

Continuity Condition

The continuity condition from Section 4.2.1 remains valid in the discrete case and is unaffected by the discretization. In the discrete case it has the form

$$\int_{(n-1+i_s)\Delta}^{H} \lambda_{(n-1+i_s)\Delta}(x)dx = v = \frac{\Delta}{\varepsilon}$$

valid for all $n \in \{1, 2, \ldots, i_b\}$. It leads again to

$$v_{(n-1+i_s)\Delta}(j\Delta) = \int_{j\Delta}^{H} \lambda_{(n-1+i_s)\Delta}(y)dy \tag{4.4}$$

valid for all $n \in \{1, 2, \ldots, i_b\}$ and $j \in \{n + i_s, \ldots, N\}$ and is used to derive the transition probabilities below.

Transition Probabilities

The probabilities for these three transitions can be deduced using the rate function $\lambda_h(x)$ from Equation (4.1). The rate of particles moving from a cell into the annular region around the slug and being deposited directly underneath the slug at cell n is given as the average probability rate for removal from that cell multiplied by one time unit ε and calculated below. The upward movement can be derived with Equation (4.4). We obtain

$$p_{i,n}(n) = \varepsilon \frac{1}{\Delta} \int_{(i-1)\Delta}^{i\Delta} \lambda_{(n-1+i_s)\Delta}(x) dx \tag{4.5}$$

$$p_{i,i+1}(n) = \varepsilon \frac{1}{\Delta} \int_{i\Delta}^{H} \lambda_{(n-1+i_s)\Delta}(x) dx \tag{4.6}$$

$$p_{i,i}(n) = 1 - p_{i,n}(n) - p_{i,i+1}(n) \tag{4.7}$$

for all $1 \leq n \leq i_b$ and $n + i_s \leq i \leq N$. All other transitions have probability 0 except for $p(n)_{i,i}$ for $i \in \{1, \ldots, n + i_s - 1\}$; these are set to equal 1. Thus the transition matrices $P(n) = (p_{ij}(n))_{1 \leq i,j \leq N}$ are defined for all $1 \leq n \leq i_b$.

The probability distribution at some time $0 \leq n_0 \leq i_b$ is now given by Equation (1.3) as

$$p(n_0) = \bar{\pi} P(1) \cdot \ldots \cdot P(n_0), \tag{4.8}$$

where $p(n_0) = (P(X_{n_0} = 1), \ldots, P(X_{n_0} = N))$. Considering $n_0 = i_b$ captures the total effect of one slug formation and rise. The probability distribution at time $n_0 = i_b$ can be interpreted as the distribution of an infinite number of particles with initial distribution π after one slug.

Segregation

What we have modeled so far is the particle behavior in a bed containing particles of uniform properties. A simple segregation process is introduced based on the idea of jetsam particles (the denser particles that tend to sink in the mixture) sinking when slugs disturb the system. Segregation of jetsam and flotsam particles in a slugging fluidized bed manifests itself as a difference in the downward drifts. The mass balance equation (4.1) only holds on average with $\int_h^H \lambda_h(x) dx > v$ for jetsam particles and $\int_h^H \lambda_h(x) dx < v$ for flotsam particles.

The segregation effect is simulated by increasing the probability of a down-transition for the jetsam particles giving an extra downward drift. A simple way to account for larger downward flow of the jetsam particles is to modify the transition probabilities (4.5), (4.6) and (4.7) as follows:

$$\bar{p}_{i,n}(n) = \alpha + (1-\alpha)p_{i,n}(n), \qquad (4.9)$$

$$\bar{p}_{i,i+1}(n) = (1-\alpha)p_{i,i+1}(n), \qquad (4.10)$$

$$\bar{p}_{i,i}(n) = (1-\alpha)p_{i,i}(n) \qquad (4.11)$$

for all $1 \leq n \leq i_b$ and $n + i_s \leq i \leq N$. Here $0 \leq \alpha \leq 1$ denotes a constant defining the strength of the segregation effects. When calculating the probability distributions for jetsam particles the transition probabilites (4.9), (4.10) and (4.11) have to be used in Equation (4.8).

Superposition

Up to now only the time window $\{0, 1, \ldots, i_b\}$ is fully investigated. We extend the definition of the transition probabilities to the full index set $\mathbb{N}$ of the Markov chain by setting

$$p(n)_{i,j} = p(n \bmod i_b)_{i,j}$$

for all $n \in \mathbb{N}$ and $i, j \in \{1, 2, \ldots, N\}$. Hence the Markov chain $(X_n)_{n \in \mathbb{N}}$ is well defined on $\{1, 2, \ldots, N\}$.

The displacement of a particle caused by consecutive slugs is regarded as a superposition by several slugs. This yields for a number of slugs $n_s \in \mathbb{N}$ that the probability distribution of the particle's position given its initial probability distribution π after some time $m = n_s i_b$ is given by

$$(P(X_m = i))_{1 \leq i \leq N} = \pi(SP(1) \cdot \ldots \cdot P(i_b))^{n_s}, \qquad (4.12)$$

with X_m denoting the jetsam particle's position after n_s slugs, $P(i)$ for $1 \leq i \leq i_b$ defined by Equations (4.9), (4.10), (4.11) and the shift matrix S defined by Equation (4.3).

4.3 Comparison of Model and Experiment

An application of the Markov chain model for the movement of a particle in a slugging fluidized bed reactor that is presented in the previous sections to two different sets of experimental data is topic in the current section. It is shown that the model catches the essential features and characteristics of the behavior of slugging fluidized beds of two different empirical beds.

4.3.1 Experimental Setup

Two sets of experimental data are compared to results from the model. They are from system which show a different behavior. The first set is from the article by Abanades and Atarés (1998) while the second set can be found in Abanades, Kelly and Reed (1994).

Abanades and Atarés (1998)

Abanades and Atarés (1998) arranged a layer of white jetsam particles in a bed of flotsam red particles. The particles in these experiments were composed out of pigments agglomerates. The mixture was suddenly fluidised at a pre–set gas velocity and the process recorded on videotape. The experiments were carried out in a column of 15 cm diameter and were performed at 3 superficial gas velocities, 1.57, 1.69 and 1.83 m/s, and 2 initial bed heights, 0.53 m and 0.73 m. For more details we refer to Abanades and Atarés (1998).

Abanades, Kelly and Reed (1994)

In 1994 Abanades, Kelly and Reed investigated a slugging bed containing coal and limestone; a completely different system than the one in Abanades and Atarés (1998). Coal acts as a flotsam and limestone as jetsam. The experiments were carried out in a column of 13 cm diameter and were performed at 2 superficial gas velocities, 0.8 m/s and 0.95 m/s, and 2 initial bed height 0.75 m. Details can be found in Abanades, Kelly and Reed (1994).

4.3.2 Model Setup

To setup the model we need to fix the discretization size Δ, determine the heights of slug and bed at rest, the slug frequency, the segretion constant α and the velocity of the slug and convert the volumetric flow between the cells to transition probabilities.

The transition probabilities can only be derived if information about the density functions λ_h is know. We assume that the density functions follow a power law with some parameter $r \geq 1$, i.e.

$$\lambda_h(x) = c(H + h - x)^r$$

holds for all $h \in (h-s, H)$ and $x \in [h, H]$ and some $c \in \mathbb{R}$. The constant c is determined by the continuity condition (4.1). This yields for the density functions

$$\lambda_h(x) = \frac{v(r+1)}{H^{r+1} - h^{r+1}}(H + h - x)^r, \qquad (4.13)$$

for all $h_s \leq h \leq H$ and $h \leq x \leq H$. The nontrivial transition probabilities without segregation can now be calculated with Equations (4.5), (4.6) and (4.7) as

$$\begin{aligned}
p_{i,n}(n) &= \varepsilon \frac{1}{\Delta} \int_{(i-1)\Delta}^{i\Delta} \lambda_{(n-1+i_s)\Delta}(x) dx \\
&= \frac{(H+(n-i+i_s)\Delta)^{r+1} - (H+(n-i-1+i_s)\Delta)^{r+1}}{H^{r+1} - ((n-1+i_s)\Delta)^{r+1}}
\end{aligned} \qquad (4.14)$$

$$\begin{aligned}
p_{i,i+1}(n) &= \varepsilon \frac{1}{\Delta} \int_{i\Delta}^{H} \lambda_{(n-1+i_s)\Delta}(x) dx \\
&= \frac{(H+(n-i-1+i_s)\Delta)^{r+1} - ((n-1+i_s)\Delta)^{r+1}}{H^{r+1} - ((n-1+i_s)\Delta)^{r+1}}
\end{aligned} \qquad (4.15)$$

$$\begin{aligned}
p_{i,i}(n) &= 1 - p_{i,n}(n) - p_{i,i+1}(n) \\
&= \varepsilon \frac{1}{\Delta} \int_{(n-1+i_s)\Delta}^{(i-1)\Delta} \lambda_{(n-1+i_s)\Delta}(x) dx = \frac{H^{r+1} - (H+(n-i+i_s)\Delta)^{r+1}}{H^{r+1} - ((n-1+i_s)\Delta)^{r+1}}
\end{aligned} \qquad (4.16)$$

for all $1 \leq n \leq i_b$ and $n + i_s \leq i \leq N$. After including the segregation effect by using Equations (4.9), (4.10) and (4.11) the final transition probabilites are found as functions of the segregation parameter α and the rate parameter r. Their choice is stated in Figures 4.5 and 4.6.

The model predictions for Figures 4.5 and 4.6 in the next section were calculated using MATLAB. Equation (4.12) was implemented with number of cells N between 15 and 17 ($N - 10$ of these were occupied by the slug), diameter of column $D = 0.15\ m$ respectively $D = 0.13\ m$, minimum fluidization velocity $u_{mf} = 1\ m/s$ respectively $u_{mf} = 0.16\ m/s$ and height of bed h_b, segregation parameter α, parameter for the flow into the annulus region r, superficial gas velocity u, time t as given in the figures. The total tracer volume was given via the total tracer volume in the experimental data. It was assumed that slugging begins on the distributor plate and the slug frequency f can, following Abanades and Atarés (1998), be calculated from the empirical correlation:

$$f = 0.32 \frac{u^{-0.15}}{h_b}.$$

All the physical properties were based on the axisymmetrical slug type. The height of a stable slug h_s was calculated via solving

$$\frac{h_s}{D} - 0.495 \sqrt{\frac{h_s}{D}} \left(1 - \frac{u - u_{mf}}{0.35\sqrt{gD}}\right) + 0.061 - \frac{1.939(u - u_{mf})}{0.35\sqrt{gD}} = 0$$

following Davidson, Clift and Harrison (1985). The calculations took only a few seconds on a standard PC.

4.3.3 Results

We compare the model results with the actual data in Figures 4.5 and 4.6. The parameters α and r are chosen by hand to achieve decent fits since theory to determine them still lacks. The two sets of experimental data are discussed separately analogous to the previous section.

Abanades and Atarés (1998)

Comparing the data from Abanades and Atarés (1998) with the model results in Figure 4.5 shows good agreement between the predicted and experimental data. There is one weak effect in the experimental data that is not accounted for in the model predictions. The tracer fraction in the upper part of the bed increases slightly with bed height when in the model predictions the tracer fraction is a decreasing function of bed height. It should be possible to achieve even better predictions with more knowledge on how to choose r and α. Here the parameters are chosen as $\alpha = 0.17$ and $r = 1$.

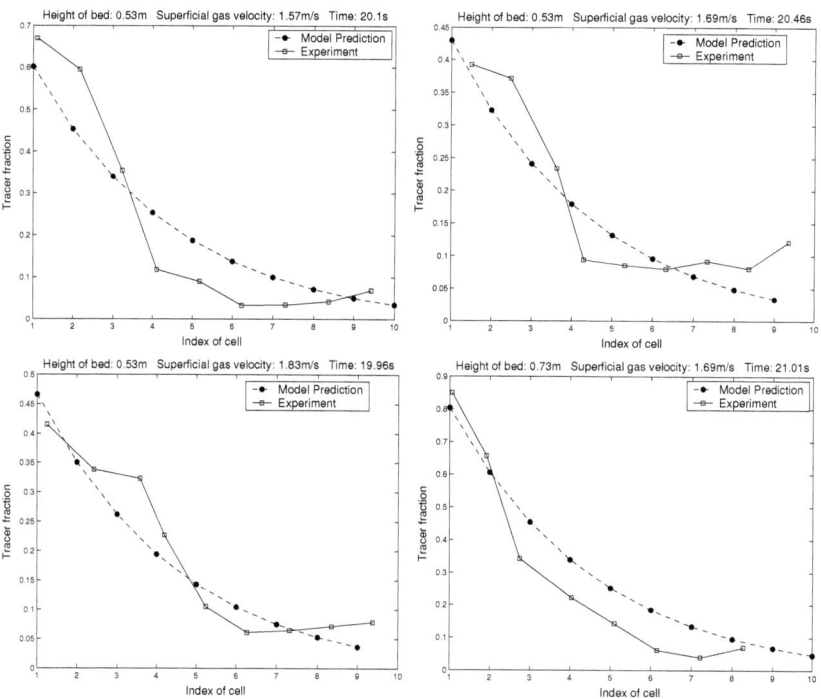

Figure 4.5: Distribution of jetsam volume fraction with $\alpha = 0.17$ and $r = 1$. (Data from Abanades and Atarés (1998))

Abanades, Kelly and Reed (1994)

The parameters $\alpha = 0.07$ and $r = 1$ provide good results for the model associated to the experimental data from Abanades, Kelly and Reed (1994) as can be seen in Figure 4.6. Again good agreement between experimental data and model predicition is detected although this system differs significantly from the one in Abanades and Atarés (1998).

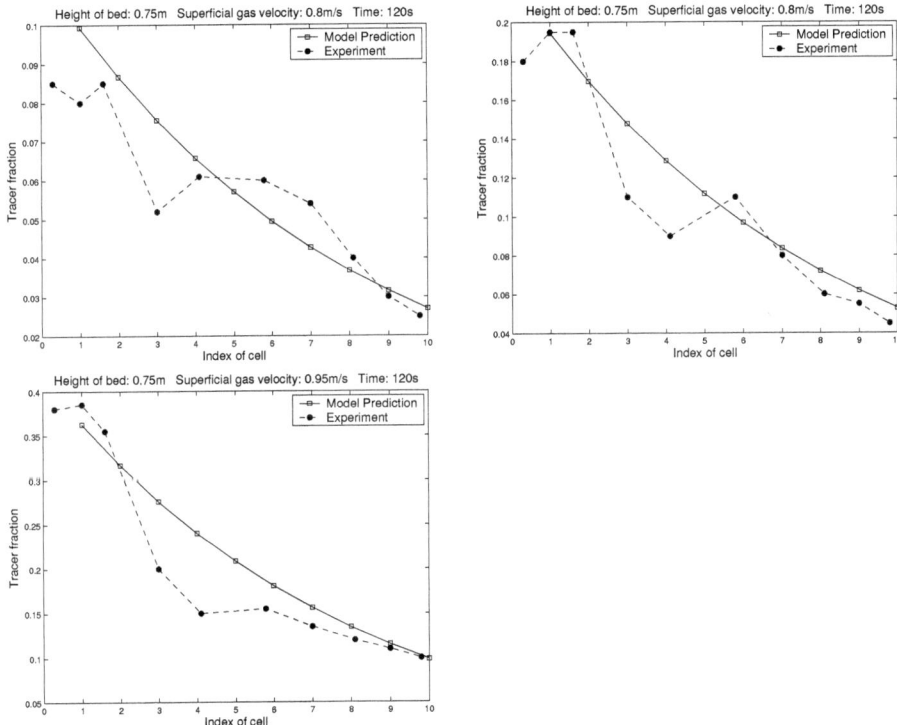

Figure 4.6: Distribution of jetsam volume fraction for with $\alpha = 0.07$ and $r = 1$. (Data from Abanades, Kelly and Reed (1994)

Summary and Outlook

The presented Markov chain model for slugging fluidized beds obviously delivers good results when compared to empirical data. It constitutes a starting point from which to press ahead in the investigation of slugging fluidized beds. Its formulation is intuitive and simple but nevertheless it keeps all crucial features and characteristics of slugging fluidized beds. Its implementation and handling are easy such that it appeals to the user. It turns out to be a valuable tool from which researchers and operators who work with slugging fluidized beds can strongly benefit.

However, there are still open questions. The effects of different choices of the segregation parameters α and the rate functions λ_h are unclear. Therefore it is hard to estimate how much the given results can be improved with an optimal choice of α and λ_h. The presented model is a discrete one. For a sensible way of refining the discretization a limit should exist which represents an analogue continuous process. This link to the continuous theory is also missing up to now.

For a more thorough discussion with a stronger focus on application and engineering matters we like to refer to Dehling, Dechsiri, Gottschalk, Wright and Hoffmann (2006).

Bibliography

J.C. Abanades, S. Atarés (1998). *Investigation of solid mixing in a deep fluidized bed of coarse particles by image analysis*, Proceedings of the World Conference on Particle Technology 3, **221**.

J.C. Abanades, S. Kelly, G.P. Reed (1994). *A mathematical model for segregation of limestone-coal mixtures in slugging fluidized beds*, Chemical Engineering Science, **43**, 3943-3953.

J. Baeyens, D. Geldart (1974). *An investigation into slugging fluidized beds*, Chemical Engineering Science, **29**, 255-265.

H. Bauer (1996). *Probability Theory*, de Gruyter.

R.N. Bhattacharya, E.C. Waymire (1990). *Stochastic Processes with Applications*, John Wiley & Sons.

A.R. Dabrowski, H.G. Dehling (1998). *Jump diffusion approximation for a Markovian transport model*, in Asymptotic Methods in Probability and Statistics, Szyszkowicz B. (editor), Elsevier Science, North-Holland, Amsterdam, 115-125.

P.V. Dankwerts (1953). *Continuous flow systems*, Chemical Engineering Science, **2**, 1-13.

J.F. Davidson, R. Clift, D. Harrison (1985). *Fluidization*, 2nd edition, Academic Press, New York.

J.F. Davidson, D. Harrison (1971). *Fluidization*, Academic Press, New York, chapter 5.

C. Dechsiri (2004). *Particle Transport in Fluidized Beds*, Ph.D. thesis, Groningen.

C. Dechsiri, E.A. van der Zwan, H.G. Dehling, A.C. Hoffmann (2004). *Positron emission tomography applied to fluidization engineering*, Canadian Journal of Chemical Engineering, **83**, 88-96.

H.G. Dehling, C. Dechsiri, T. Gottschalk, P.C. Wright, A.C. Hoffmann (2006). *A stochastic model for mixing and segregation in slugging fluidized beds*, Powder Technology, **171**, 118-125.

H.G. Dehling, T. Gottschalk, A.C. Hoffmann (2007). *Stochastic Modeling in Process Technology*, Elsevier.

H.G. Dehling, A.C. Hoffmann, H.W. Stuut (1999). *Stochastic models for transport in a fluidized bed*, SIAM Journal on Applied Mathematics, **60**, 337-358.

R.M. Dudley (2002). *Real Analysis and Probability*, Cambridge University Press.

A. Einstein (1905). *Über die von der molekularkinetischen Theorie der Wärme geforderte Bewegung von in ruhenden Flüssigkeiten suspendierten Teilchen*, Annalen der Physik, **17**, 549-560.

K.-J. Engel, R. Nagel (2000). *One-parameter Semigroups for Linear Evolution Equations*, Springer-Verlag.

W. Feller (1952). *The parabolic differential equations and the associated semi-groups of transformations*, Annals of Mathematics, **55**, 468-519.

W. Feller (1954). *Diffusion processes in one dimension*, Transactions of the American Mathematical Society, **97**, 1-31.

W. Feller (1971). *An Introduction to Probability Theory and its Applications*, Vol. II, 2nd ed., John Wiley & Sons.

L.G. Gibilaro (1979). *Residence time distributions in regions of continuous flow systems*, Chemical Engineering Science, **34**, 697-702.

T. Gottschalk, H.G. Dehling, A.C. Hoffmann (2006). *Danckwerts' law for mean residence time revisited*, Chemical Engineering Science, **61**, 6213-6217.

C.M. Grinstead, J.L. Snell (1997). *Introduction to Probability*, 2nd rev. ed., American Mathematical Society, chapter 11.2.

A.C. Hoffmann, H. Paarhuis (1990). *A study of the particle residence time distribution in continuous fluidized beds*, I. Chem. E. Symposium Series, **121**, 37-49.

K. Itô (1944). *Stochastic integral*, Proceedings of the Imperial Academy Tokyo, **20**, 519-524.

U. Krengel (2002). *Einführung in die Wahrscheinlichkeitstheorie und Statistik*, Vieweg.

J. Lamperti (1977). *Stochastic Processes. A Survey of the Mathematical Theory*, Springer-Verlag.

J.M. Matsen, S. Hovmand, J.F. Davidson (1969). *Expansion of fluidized beds in slug flow*, Chemical Engineering Science, **24**, 1743-1754.

L.C.G. Rogers, D. Williams (1987). *Diffusions, Markov Processes and Martingales*, Vol. 2 Itô Calculus, John Wiley & Sons.

P.N. Rowe, B.A. Partridge (1962). *Particle movement caused by bubbles in a fluidised bed* in: Rottenburg, P. A. (hon. ed.), Interaction between fluids and particles. Institution of Chemical Engineers, London, 135-142.

J.C. Schouten, P.J.M. Valkenburg, C.M. van den Bleek (1988). *Segregation in slugging fbc large-particle system*, Powder Technology, **54**, 85-98.

A.V. Skorohod (1976). *On a Representation of Random Variables*, Theory of Probability and its Applications, **21**, 628-632.

D.W. Stroock, S.R.S. Varadhan (1979). *Multidimensional Diffusions*, Springer-Verlag, New York.

K.R. Westerterp, W.P.M. van Swaaij, A.A.C.M. Beenackers (1987). *Chemical Reactor Design and Operation*, John Wiley & Sons Ltd., Chichester, chapter IV.

N. Wiener (1923). *Differential-space*, Journal of Mathematics and Physics, **2**, 131-174.

SVH Südwestdeutscher Verlag für Hochschulschriften

Wissenschaftlicher Buchverlag bietet
kostenfreie
Publikation
von
Dissertationen und Habilitationen

Sie verfügen über eine wissenschaftliche Abschlußarbeit zu aktuellen oder zeitlosen Fragestellungen, die hohen inhaltlichen und formalen Anspruchen genügt, und haben **Interesse an einer honorarvergüteten Publikation?**

Dann senden Sie bitte erste Informationen über Ihre Arbeit per Email an: info@svh-verlag.de.

Unser Außenlektorat meldet sich umgehend bei Ihnen.

Südwestdeutscher Verlag für Hochschulschriften
Aktiengesellschaft & Co. KG
Dudweiler Landstr. 99
D – 66123 Saarbrücken
www.svh-verlag.de

Printed by Books on Demand GmbH, Norderstedt / Germany